Construction Reliability

Construction Reliability

Safety, Variability and Sustainability

Edited by
Julien Baroth
Franck Schoefs
Denys Breysse

First published 2011 in Great Britain and the United States by ISTE Ltd and John Wiley & Sons, Inc.

ISTE Ltd
27-37 St George's Road
London SW19 4EU
UK

www.iste.co.uk

John Wiley & Sons, Inc.
111 River Street
Hoboken, NJ 07030
USA

www.wiley.com

© ISTE Ltd 2011

Library of Congress Cataloging-in-Publication Data

Construction reliability / edited by Julien Baroth, Franck Schoefs, Denys Breysse.
    p. cm.
  Includes bibliographical references and index.
  ISBN 978-1-84821-230-5
  1. Buildings--Reliability. 2. Public works--Reliability. 3. Structural failures--Prevention. I. Baroth, Julien. II. Schoefs, Franck. III. Breysse, D.
  TA656.C68 2011
  624--dc23

                                                                                                2011019207

British Library Cataloguing-in-Publication Data
A CIP record for this book is available from the British Library
ISBN 978-1-84821-230-5

Printed and bound in Great Britain by CPI Antony Rowe, Chippenham and Eastbourne.

MIX
Paper from
responsible sources
FSC® C013604

Table of Contents

Preface

From 26 to 28 March 2008, a conference entitled "Fiabilité des matériaux et des structures" ("Reliability of materials and structures", JNFiab'08)[1] took place at the University of Nantes, France, bringing together the French scientific communities interested in reliability and risk analysis, as applied to materials and structures. This colloquium followed on from several different events: the fifth "Reliability of materials and structures" conference, the second Méc@proba training day[2] and the second scientific session in the subject area of "Understanding risk in civil engineering" (MRGenCi scientific interest group[3]).

It combined their themes and concerns as an extension of the first shared workshop between the Associations Françaises de Génie Civil (AFGC, or *French Associations of Civil Engineering*[4]) and the Associations Françaises de Méchanique (AFM, or *French Associations of Mechanical Engineering*)[5], during the twenty-fifth annual meeting of the Association Universitaire de Génie Civil (AUGC, *Universities civil engineering association*[6]) held on 23–25 May 2007, in Bordeaux, France.

This book was first conceived during these sessions, organized by the MRGenCi and Méc@Proba scientific interest groups, where the authors gave presentations on the advances they have made in their respective fields.

Although the examples of structures that can be found in this book fall under the umbrella of civil engineering (nuclear and oil industries, buildings and dams), the

---

1 http://www.sciences.univ-nantes.fr/jfms2008/DownloadJFMS2008.pdf.

2 http://www.lamsid.cnrs-bellevue.fr/vf/actualites/journee_proba/plaquette_proba.pdf.

3 http://www.mrgenci.u-bordeaux1.fr/.

4 http://www.afgc.asso.fr/.

5 http://www.afm.asso.fr/.

6 http://www.augc.asso.fr/.

methods we consider are just as applicable to any sort of complex mechanical system involving a large number of uncertainties. Thus the book is of interest to the civil engineering community but also to mechanical engineers or those interested in reliability theory, whether their background is in industry or academia, who have been exposed to research and development processes. Masters students, engineering students and doctoral students, engineers and research associates will all find a detailed discussion of methods and applications.

The authors are indebted to the two main proofreaders, with their complementary backgrounds. The first is Maurice Lemaire, a university professor who teaches at the Institut Français de Mécanique Avancée (IFMA, *French Institute for Advanced Mechanical Engineering*[7]) and at the Blaise Pascal University[8] (UBP) at Clermont-Ferrand, and who is consultant to the company Phimeca[9] which he co-founded. The second is André Lannoy, Vice-President of the Institut pour la Maîtrise des Risques (IMdR, *Institute for Risk Management*[10]), who built his career as a research engineer and subsequently as scientific adviser to the research and development section of EDF. In particular, André Lannoy co-organizes the working group "Sécurité et sûreté des structures" (GTR 3S, *Safety and reliability of structures*[11]), a group which counts several of the authors of this book among its members.

The authors would like to express their particular gratitude to Maurice Lemaire for his contributions to the development of the field of structural reliability. The few brief paragraphs below give a short overview of his career and his contributions to scientific production, advocacy and above all to training, which has inspired several of the authors to pursue their career paths.

After receiving his diploma from the Institut National des Sciences Appliquées, (INSA[12], *National Institute for Applied Sciences*), in Lyon in 1968, followed by further studies in applied mathematics (1969), Maurice Lemaire received his doctorate in engineering (1971). Following his higher state doctorate (1975), he was appointed a position at CUST (the engineering school that became the Polytech' Clermont-Ferrand[13]), within the Blaise Pascal University (1976). He was involved in the founding of the IFMA (1987), where he has been a professor since 1991.

---

7 http://www.ifma.fr/.

8 http://www.univ-bpclermont.fr/.

9 http://www.phimeca.com/.

10 http://www.imdr.eu.

11 http://www.imdr.eu/v2/extranet/detail_gtr.php?id=36.

12 http://www.insa-france.fr/.

13 http://www.polytech-clermontferrand.fr/.

Maurice Lemaire founded the Laboratoire de Recherches et applications en Mécanique Avancée (LaRAMA, *Research and applications in advanced mechanics* research group) in 1989 (and the IFMA/UBP laboratory, which is now LaMI[14]), where he has been a director since 2005. Head of research at IFMA from 1991 to 2007, he has supervised 44 doctoral students and participated in 180 thesis examinations.

Co-founder of the "Fiabilité des matériaux et des structures" symposium[15] (Cachan, 1994), co-organizer of the 7th *International Conference on Applications of Statistics and Probability in Civil Engineering* (ICASP[16], 1995 in Paris), and then president of the *International Civil Engineering Risk and Reliability Association* (CERRA[17]) from 1995 to 1999, he is a member of the scientific committee of the International Federation for Information Processing's working group 7.5 (IFIP WG 7.5) on the reliability and optimization of structural systems[18]. Maurice Lemaire is also a founder of the Méc@proba meetings[19] (Marne-la-Vallée, France, 2006).

He is a promoter for the scientific commission "Mécanique probabiliste des matériaux et des structures" (*Probabilistic mechanics of materials and structures*[20]) of the AFM), and is responsible for numerous industrial research projects. He also contributes scientific insight and ongoing supervision into the growth of the company Phimeca Engineering, which he co-founded in 2001.

There can be no doubt that Maurice Lemaire has contributed to the development of the science of structural reliability at a national and international level.

The authors therefore recommend that readers consult his publications, which are cited throughout this book.

The authors would like to express their gratitude towards the MRGenCi scientific interest group which made it possible to write this book. That group, created on the initiative of Messrs Breysse (Professeur at University of Bordeaux I), Boissier (Professeur at UBP), Melacca (SMA-BTP) and Gérard (PDG OXAND SA), now consists of more than 30 industrial companies, learned societies, universities and engineering schools.

---

14 www.ifma.fr/lami/.

15 www.sciences.univ-nantes.fr/jfms2008/DownloadJFMS2008.pdf.

16 www.icassp2011.com/.

17 www.ce.berkeley.edu/ projects/cerra/.

18 www.era.bv.tum.de/IFIP/.

19 www.lamsid.cnrs-bellevue.fr/vf/actualites/journee_proba/plaquette_proba.pdf.

20 www.afm.asso.fr/PrésentationdelAFM/Commissions/MPMS.

We cannot speak highly enough of the authors, proofreaders and all the other people who have contributed to this book and to ensuring that it is received by a wide audience.

Julien BAROTH
Franck SCHOEFS
Denys BREYSSE
JUNE 2011

Introduction

Background and objectives

This book describes and illustrates methods to improve prediction of the lifetime and management of civil engineering structures. This contributed collection aims to complete the existing literature and to provide access to both recent scientific approaches and examples of applications in study cases. The authors are drawn from amongst university academics, senior engineers and scientific managers in companies. Amongst others, Ditlevsen & Madsen [DIT 96], Melchers [MEL 99] and Lemaire [LEM 09] have already made significant contributions to structural reliability, that is to say, the study of the ability of a structure to perform a function (depending on its environment, life, etc.). Favre [FAV 04], [SUD 08b], by contrast, studied geotechnical structures. These last books introduced concepts of uncertainty related to materials and loads, as well as statistical and probabilistic methods applied to civil engineering. This book also uses these basic concepts, and notions of uncertainty or reliability, in particular, are discussed in this introduction.

The authors' main objective in this book is the presentation of recent methods of data processing and computation which have not been widely disseminated. Many of these methods have already been used in industrial applications: this book aims to present these applications through case studies and examples over a wide set of materials, buildings and structures in civil engineering.

If the examples of structures presented in the following pages are limited to the field of civil engineering (whether in nuclear, oil, or dam building applications), the methods presented are applicable to any complex mechanical system in an uncertain environment. This book is mainly intended for those familiar with civil engineering, engineering mechanics or the theory of reliability, both from industry and academia,

Introduction written by Julien BAROTH, Alaa CHATEAUNEUF and Franck SCHOEFS.

involved in research and development. Students, especially engineering students and PhD students, engineers and research fellows may also have an interest in the book.

A civil engineering project is considered here to be a system that ensures one or more global functions. It consists of components, sub-structures, structures, human actors, procedures, and organization in a given environment. It performs the basic functions that contribute to the achievement of these global functions. The system is identifiable, and can be broken down into the functions it performs (a functional approach), or into structurally interdependent elements or subsystems (a structural approach). A civil engineering project is considered safe when it is fit for duty for the duration of its service life, without damage to itself or to its environment (French norm X60-010). The safety of a structure can be quantified by its probability of not providing any of its functions, at any moment of its expected lifetime.

This probability of failure, denoted P_f, is mathematically complementary to the system reliability, often denoted $R = 1-P_f$.

The objective of this book is to present the methods of functional analysis (Parts 1, 4 and 5) and those of structural analysis (throughout the entire book), when the systems studied are in a working condition that does not induce damage to people, property or the environment. These systems are then considered to be safe. A significant section of the book is devoted to the calculation of reliability, which may or may not be dependent on time (see Parts 3 and 4).

Some of the concepts developed throughout the book are introduced in the following section, which also raises a number of questions that the book later tries to answer and illustrate.

Qualitative and quantitative methods of safety assessment for a construction project

The study of the safety of a civil engineering structure may be conducted by system analysis, followed by an analysis of failure modes and modeling of failure scenarios, in order to answer some crucial questions:

Question 1. How can we highlight the most likely failures and the most critical failure scenarios, which could potentially be the basis of risk analysis?

Question 2. What are the preventive measures that can improve system safety?

Part 1 of this book addresses these issues. It presents some methods of qualitative assessment of structural safety. These methods allow us to analyze a system, its failure modes, and to model the failure scenarios in order to evaluate their criticality. An application of methods to assess the criticality of scenarios is then proposed for a hydro civil engineering work.

Moving beyond qualitative methods, quantitative methods are also presented in Parts 2 to 5.

For a given failure scenario, a third question might be:

Question 3. How can we evaluate and reduce, when possible, the probability of events causing failure or, again when possible, how can we evaluate and reduce the gravity of failure consequences?

Some answers to this question are given by an analysis of uncertainties in knowledge of a structure and the actions applied to it. It has been a deliberate choice in this book to focus on early uncertainties, in particular focusing on uncertainties in our knowledge of building materials. With regard to the study of applied actions, readers can refer to the following sources: [CEN 03], [CEN 05] on snow and wind climate loads, [SCH 08], [SMI 96] for wave action, or [PEY 09] regarding loads from hydrology.

Materials with uncertain properties

There are many sources of uncertainty related to materials. Being aware of them, or reducing them, will achieve greater safety [FAV 04]. Thus, the whole book attempts to answer the following questions:

Question 4. How can we represent and use uncertain data, describing the geotechnical characteristics of materials? What are the consequences of heterogeneity and variability for structural safety?

A first classification consists of combining uncertainties into two categories concerning the condition of the work considered:

– uncertainties related to internal variables affecting the internal condition of the structure, such as material properties (elastic modulus, Poisson's ratio, density, coefficient of thermal expansion, etc.), geometrical parameters (dimensions, moments of inertia, etc.), internal boundary conditions (excluding actions on areas or volumes) and internal links;

– uncertainties related to external variables beyond the internal condition of the structure, such as natural actions (wind, waves, snow, temperature, earthquake, etc.) and operating loads (loads, operating loads, etc.).

In some cases these differences may be small, e.g. for a metal sheet pile wall, where uncertainty about the weight of the soil near the wall is external if we consider the resistance of the curtain, and internal if we consider stability (sliding). Table I.1 provides a more detailed typology of uncertainties that are not problematic.

Type of uncertainty		Nature of uncertainty	Way to reduce uncertainty
Aleatoric	Inherent uncertainties (natural)	Environmental parameters, spatial or temporal variability of material properties	Cannot be reduced or eliminated; can only be quantified (development of knowledge, data acquisition) and taken into account[1]
Epistemic	Model uncertainties: Physical model	Empirical and theoretical relationships to describe physical processes (adequacy between model and reality)	Can be reduced by improving knowledge and models
	Model uncertainties: Statistical distribution	Related to the limited data available (sampling), method of collection, etc.	Can be reduced by treating the collection of data (more numerous, more accurate)
Ontologic	Human and organizational errors	Human error and organizational operators, procedures, equipment and coordination among stakeholders	Can be reduced through better organization and increased skills (monitoring, quality, training)

[1] *A better quality of processes and control scans reduce variability and remove glaring defects.*

Table I.1. *Typology of uncertainties and means of action*

A second classification distinguishes random (or intrinsic) uncertainties and epistemic uncertainties, which are sometimes measurable and are likely to be modeled by random variables. This probabilistic format is easier to introduce into behavior models and to propagate through structural analyses. Amongst the uncertainties, we encounter the physical uncertainties that are inherent to the nature of a system, such as uncertain parameters: strength, loads, environment, geometric characteristics, boundary conditions, etc. However, uncertainties due to lack of knowledge of physical phenomena are epistemic because they can be reduced by getting more information, or by doing further research.

However, the gap between a real system and a model, whether simple or complicated, can also lead to uncertainty. This type of model uncertainty is epistemic: it can sometimes be incorporated into a calculation of reliability through the introduction of additional random variables to represent the dispersion of the model results (analytical, numerical or experimental) compared to the physical phenomenon.

Table I.1 presents more ontological (or phenomenological) uncertainties, harder to identify than others. It describes the uncertainties attached to random processes or systems which have not been imagined (for example, the Concorde crash in Paris in 2000) or which are too complex to be modeled [TAN 07].

Engineers should adopt an alternative attitude, just as philosophers should avoid the question: is reality random or not? Engineers are simply interested in objects, processes and phenomena that are not entirely predictable: uncertainty concerns engineers to the extent that it limits their ability to forecast [MAT 08]. More than the classification of uncertainties (which depends on a refinement of the models used), the priority is how to reduce or control them (as raised in Questions 3 and 4).

Eurocode principles

The European building code (Eurocodes) has been developed over the last 20 years. This code allows the variability of materials in a deterministic formalism to be better accounted for, based on values defined on a probabilistic basis. This formalism is also based on failure situations formalized according to their consequences in terms of Serviceability Limit States (SLS) and Ultimate Limit States (ULS). SLS correspond to conditions beyond which the serviceability requirements specified for the work (or a part of it) are not met. These statements are generally related to demands of comfort (vibration, deflection), aesthetics (excessive distortion), durability (cracking, corrosion), the proper functioning of equipment (insulation, sealing, etc.), without resulting in short-term collapse of a structure. ULS are associated with the collapse of a structure, or other forms of structural failure: loosing the equilibrium of the work, reaching the maximum strength of a part of the structure, failure of the subgrade; it is defined as the accidental limit state (or progressive failure) for systems exposed to variable loads and to a fatigue limit state (condition of sustainability). The principle of the Eurocodes can be summarized as:

– proposing an appropriate limit state for each failure scenario;

– characterizing the parameters affecting the considered limit state criteria, using probabilistic and statistical tools;

– replacing the complete distributions of the probabilistic variables by average values and dispersion or, by characteristics values;

– neglecting the dispersion of some data, considering them as deterministic;

– taking into account the other neglected uncertainties, by introducing fixed coefficients.

This regulatory approach is characterized as semi-probabilistic, because it incorporates a probabilistic modeling of uncertain parameters (b), while introducing partial safety factors (e). The second part of the Code recalls the definitions of particular values (characteristics values), (c), and uses the concept of semi-probabilistic design. The Eurocodes are a compromise between inadequate deterministic design, and impractical probabilistic design.

We can also make a distinction between four levels of reliability methods:

– Level I: methods using some specific values for the variables (resistance and load, for example); each random variable is represented by a characteristic value with statistical content which is often poorly defined;

– Level II: methods devoted to characterizing the random variables by their mean and variance;

– Level III: methods requiring more knowledge of the joint probability distribution of all the random variables, and allowing a reliability index and failure probability to be obtained as a measure of safety;

– Level IV: methods intending to provide a level of reliability integrating economic criteria: for example, taking into account the costs of construction, maintenance, repairs, failure consequences, etc.

One or other of these four levels of reliability methods may be considered, depending on the size of the problem to be analyzed.

For example, Level IV methods are devoted to the analysis of nuclear power plants, whilst Level I methods are applied when studying simple structures with lower stakes (warehouse storage products that are safe, etc.). The reliability assessment of structures with high stakes is theoretically Level III. However, a simplification is necessary, because the joint probability distribution of variables is difficult to assess. We usually just know the marginal distributions, sometimes with correlation coefficients to model the interdependence between variables. Such methods are called advanced Level II methods.

Variability and heterogeneity of materials

Part 2 of this book seeks to answer Question 3 by showing how to use available data to describe their heterogeneity and variability. Chapter 4 deals with the characterization of uncertainty in geotechnical data. This part provides a complete set of methods: the identification of sources of uncertainty described above, the classification of data (outliers, censored and poor data), and statistical representation and modeling of these data (possibilistic or probabilistic). Readers are referred to other publications on this topic including [CEL 06], [LEM 99], where answers to Question 3, and to Question 5 (below) can be found.

Question 5. How can we use limited or censored data?

Chapter 5 presents estimates related to material variability (average, characteristic values), introduces geostatistical modeling tools (variograms, estimation and simulation methods) [MAR 09]. Chapter 6 provides an example of a shallow footing for which reliability is considered; the effect of soil variability on the variability of the bearing capacity and safety of the footing is studied; finally, the spatial correlation is taken into account to study its influence on the safety of the linear footing.

Part 4, devoted to problems of time-variant reliability, completes Part 2 by showing how the enrichment of statistical analysis and the use of Bayesian approaches can be applied to samples of a small size (Chapter 11).

The computation of reliability-coupled mechanical and reliability models

Parts 2 to 5 of the book use the calculation of reliability indicators and answer the question:

Question 6. How can we quantify the reliability of a system or a structure?

This calculation involves various levels of complexity, concerning mechanical and probabilistic analyses. The complexity of the probabilistic model depends on the distributions of the variables, their physical limitations and their interdependence. The complexity of the mechanical behavior stems from its size (number of components), its transients and nonlinearities, etc.

When an operating system depends on its mechanical condition, there is an interdependence between the roles of mechanical and reliability variables; this is called "mechanical-reliability coupling" [LEM 00], which can be described in five stages:

– identify the purpose of the structure: its function, behavior, operating conditions and possible failures (resulting from failure analysis);

– develop predictive models of mechanical behavior with and without defects, and probability distributions of the design variables;

– identify the possible failure scenarios. The appropriate functioning of the system is defined by the performance function G_i (or limit state) to be complied with. Failure is reached if one of these limit-states is exceeded. This analysis, often overlooked, is crucial and must be as thorough as that undertaken in (b);

– for each failure scenario, determine the reliability level and the sensitivity factors. These latter are very useful in decision making for quality control and system optimization; finally,

– assess the overall failure probability of the structural system and define the partial factors to be used for the calibration of the codes and regulations.

We take x to be the vector of a model's uncertain parameters x_i, e.g. external actions (load, wind, wave, earthquake) or geometrical characteristics (size, area and moment of inertia of cross-sections), material properties (yield stress, Young's modulus, Poisson's ratio, etc.). We model each parameter with a random variable X_i, characterized by a probability distribution representing the uncertainty related to this parameter (probabilistic model). This can be done using statistical studies, physical observations, or expert advice (usually a possibilistic model in that case).

Each failure scenario is associated with a performance function (also known as a limit state function, or a safety margin), denoted G. The inequality $G > 0$ indicates the safety domain D_s, and inversely $G \leq 0$ indicates the failure domain D_f. The objective is to evaluate the probability P_f that the realizations of random variables belong to the failure domain. In the simple case of two variables representing the resistance R and load S, the performance function (or safety margin) is written as $G(R, S) = R-S$. In practice, the statistical parameters of the stress S, and sometimes those of the resistance R, are not directly accessible, because the measurements and observations are made only for the basic uncertain parameters of the vector x from which R and S originate. The variable G and the random vector X modeling x are connected by a mechanical transformation: $G(R, S) = G(X)$.

We can consider that this transformation is known (e.g. calculating a loading effect from a given depth of snow), although in some cases it is only available using an algorithm, as for example, using finite element software. In this book, the reliability indicators most often used are failure probability and reliability indices (both Cornell and Hasofer–Lind; see [LEM 09] for a detailed presentation):

– the *probability of failure* is evaluated by integrating the joint density function over the failure domain D_f, where:

$$P_f = \text{Prob}[G \leq 0]$$

– the *Cornell index* is the distance between the median point of the G margin and the point where the margin becomes zero (i.e. the failure point); this distance is measured in terms of number of standard deviations. In other words, if m_G and σ_G represent the mean and the standard deviation of G, respectively, the Cornell index β_C is written as: $\beta_C = m_G / \sigma_G$. If the margin is normally distributed, we can easily show that the failure probability is:

$$P_f = \Phi(-\beta_C)$$

where Φ is the standard normal cumulated distribution function. This expression is accurate when the distribution of G is Gaussian (a linear combination of two Gaussian variables, such as R-S, is Gaussian). If this condition is not verified, the Cornell index only gives a measurement that is no longer explicitly linked to the probability of failure. It is even less useful, as it implies assumptions such as the linearity of the margin and the normality of distributions;

– the *Hasofer–Lind reliability index* [HAS 74] is an invariant estimator of reliability. Hasofer & Lind proposed it to change physical variables into standard Gaussian variables (i.e. with zero means and unit of standard deviation) and so that they would be statistically independent. In this so-called standard space, the failure probability is written according to the failure domain $H \leq 0$ so that:

$$\varphi_n(u) \, du_1 \cdots du_n$$

where φ_n is the probability density function of the n-variate standard normal distribution. According of Hasofer–Lind's definition, the reliability index β is the minimum distance between the origin and the limit state surface in the standard space. This distance is deduced from a hyperplane tangent to the limit state function at a point P^*, called the "design point". Finding β is thus just a question of solving the following optimization problem:

$$d(u_i) = \sqrt{\sum_i u_i^2} \quad \text{with} \quad H(u_i) \leq 0$$

Reliability methods

Different approaches are possible for coupling mechanical and reliability methods, each offering various levels of compromise between accuracy, cost and reliability indicators, with regard to the range of validity of the methods (strongly or weakly nonlinear (y=x^1.1 is weakly nonlinear, y=x^5 is strongly nonlinear) mechanical calculations, or a more or less reduced number of random variables, etc.).

Two classes of conventional methods cover the majority of current developments and applications: the Monte Carlo method and First/Second Order Reliability Methods (FORM/SORM approximations). These are introduced and used in Parts 2 to 5.

Monte Carlo simulations are the most robust method to assess the failure probability of a complex system. They enable the achievement of reference results and control other types of approximation. However, they are often expensive. In general, to properly assess probability of the order of 10^{-n}, we must perform 10^{n+2} to 10^{n+3} mechanical calculations. It is obvious that this method is impossible to use for large systems with a low probability of failure. More efficient techniques such as modified Latin Hypercube, used in [SCH 08], can be considered as alternatives.

In a standard space, the FORM/SORM methods are based on the evaluation of a *reliability index*, denoted β, followed by an approximation of the probability of failure. The Hasofer–Lind Index [HAS 74] is the most commonly used. The search for the design point P introduced earlier can be performed by an optimization method appropriately chosen with respect to the particular form of the problem.

A first approximation of P_f is obtained by replacing the boundary condition $H(u_i) = 0$ by a tangent hyperplane at the design point; this approximation is known as the First Order Reliability Method (FORM). Taking into account the rotational symmetry property of the standard probability density, we can estimate this probability by [DIT 96]:

$$P_f \approx \Phi(-\beta)$$

where Φ is the standard cumulative distribution function. The accuracy of this approximation depends on the nonlinearity of the limit state, especially in the vicinity of P^*.

The purpose of Part 3 of the book is to present another class of methods for calculating reliability, called "response surfaces", since the mechanical response, usually not explicit, is approximated by a meta-model, which is often reduced to an explicit analytical polynomial function. This group of methods has been the subject of recent developments, which are applied to examples of a truss, then to the skeleton of a building on several floors. Therefore, Part 3 answers the question:

Question 7. How do we evaluate the reliability of very expensive mechanical models?

Time variant reliability methods

Part 4 outlines the problems of time-dependent reliability using a number of different methods. The topics discussed include answers to the following questions:

Question 8. How do we implement a calculation of time-dependent reliability?

Question 9. How do we use additional information gained over time to update reliability calculations?

Modeling time-dependent variability is first done using methods of aggregation and unification of data (Chapter 9). These approaches are applied to assess the time taken to evacuate a building during a fire.

Time-dependent problems can also be analyzed using probabilistic concepts such as stochastic processes, as is the case for the study of climate change and the evolution of spatial characteristics of components. These processes can be defined by an infinite number of random variables indexed on time. In addition to the mean and standard deviation which can be time-variant, stochastic processes are characterized by auto-correlation, which implies certain dependence between neighboring points (in time) of the same process. In particular, in the problem of safety margins discussed above, R and S are modeled by stochastic processes and denoted $R(t)$ and $S(t)$. At time t during the life of a structure, the instantaneous probability of failure is:

$$P_f(t) = P[R(t) \leq S(t)]$$

If the instantaneous densities $f_R(x, t)$ and $f_S(x, t)$ are known, the instantaneous probability of failure $P_f(t)$ can be calculated by the integral:

$$F_R(x, t) \, f_S(x, t) \, dx$$

In practice, it is not possible to detect failure continuously on the time axis. If the overload is not applied in increments, we can discretize the time into small intervals in which the stress level is considered as constant, which can then approximate the instantaneous probability of failure.

Chapter 10 (especially Figure 10.5) provides an example of how to calculate instantaneous probability, not to be confused with the cumulative probability of the form:

$$P_{f,c}(t) = P\left[\exists t \in [0,t] \,/\, G(t) \le 0\right]$$

Another approach is based on the consideration of the evolution of a safety margin $G(t)$ over the lifetime of a structure. In this case, the margin is modeled by a stochastic process. We therefore seek the probability that the margin is negative or zero over the observation interval. This approach is called the method of zero-crossing. The time at which the margin becomes negative or zero for the first time is called time-to-failure, which is a random variable. The corresponding probability $P[G(t) \le 0]$ is called the probability of first zero-crossing; for non-repairable systems, it is called the probability of failure. The corresponding distribution is called the distribution of lifetime.

Methods, based on stochastic processes, exist to model degradation with or without monotonic loading, and they are introduced in Chapter 10. They can make demands in line with the actual physical/habitat conditions of the structure. Theoretically, component failure can occur at any time after its commissioning. Methods of stochastic processes [SHI 91] can reproduce loads that comply with the best environmental conditions of usual structures. These phenomena can be statistically described from measurements taken for the process over time (e.g. measurements of temperature, relative humidity, maximum stress, etc.).

The recently developed "PHI2" method [AND 04], [SUD 08a] (see Chapter 10), as well as the Monte Carlo Markov Chains (Chapter 11) are then presented. The main applications in this part of the book are a serial system (for both poor and censored data), a truss structure, and a containment building in a nuclear power plant (Chapter 12).

Resistance can sometimes be assumed to be invariant over time, but it remains dependent on the weather in general, in the case of corrosion processes e.g. [BOÉ 09]. It is also possible to adopt a pessimistic assumption associated with the distribution of extreme values (i.e. minimum) of resistance.

Lifecycle optimization

Part 5, the final part of the book, describes maintenance optimization using reliability methods. It presents the concepts of maintenance and lifecycle costs for a system. The cost models of maintenance for components and systems are defined in order to allow an optimal maintenance policy to be selected. The applications in this part of the book cover several issues related to corrosion of reinforced concrete (pre-stressed beam, cooling tower, etc.). The following paragraphs (below) present some background on the analysis of the lifecycle of construction works.

During the lifecycle of a construction project, the failure rate follows the bathtub curve, shown in Figure I.1. After a phase of early failure due to design and construction errors, the failure rate is almost constant for a large part of the structural life, as the degradation mechanisms have not yet emerged. After some time, the phenomenon of degradation starts, leading to increasing rates of failure; it is during this stage that preventive maintenance can improve structural reliability and extend the service life.

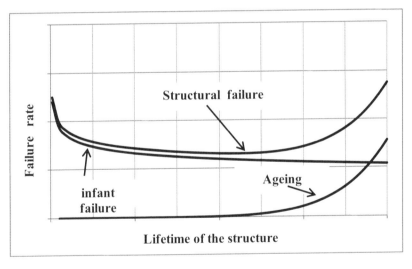

Figure I.1. *Evolution of the failure probability of a civil engineering work*

The optimization of a structure must take into account all the costs at stake throughout the structure's lifetime.

Management of service-life must take into account the level of reliability based on structural age, costs associated with various potential events (loss of function, failure, inspection, repair, etc.), the benefits given to society by the existence of the structure, and its interaction with the environment in the context of sustainable development.

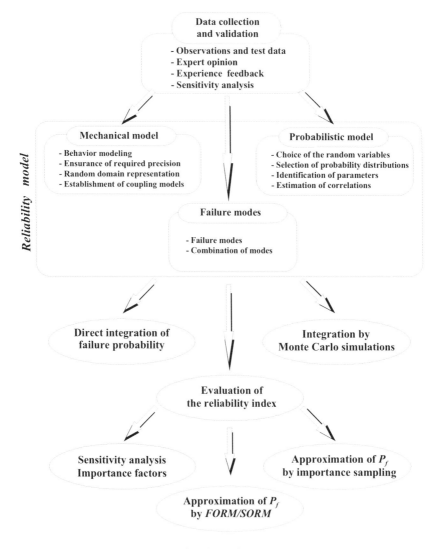

Figure I.2. *Flowchart of reliability analysis*

Given that the total cost is highly affected by uncertainties, the optimization of design and maintenance must take into account the random dimension of the total cost, in order to ensure the necessary margins and to maximize the expectation of socio-economic utility. While design consists of optimizing the distribution of materials in a structural system, maintenance seeks to optimize the methods and schedules for inspection and repair, depending on the state of degradation (estimated or observed) and the interactions between components in the structural systems.

Conclusion

This introductory chapter presented the framework and overall organization of the book. It addresses the qualitative methods for analyzing failure modes and their criticality (Part 1), and the quantitative methods of reliability (Parts 2 to 5). The notion of uncertainty related to the data of building materials is introduced and applied (Parts 2 and 4). Then, reliability methods without and then with a consideration of time scale are introduced (Parts 3 and 4, respectively). Finally, and throughout the book, an overview of system reliability is presented, first in a functional study of a dam (Part 1), second by including the study of the life cycle of civil engineering works in Part 5 (with studies of the maintenance of a stretch of motorway and of a cooling tower).

Figure I.2 provides a summary of the principles of a reliability analysis. The first step is to analyze the mechanical system, using functional and structural approaches. This first step includes the collection of available information on the system and the identification of possible failure scenarios and their criticality. A second step is to define the mechanical and probabilistic (or possibilistic) models for analysis. These two models are completed by the definition of the possible failure scenarios, to make up a coupled mechanical reliability model. In the absence of the direct integration of failure probability (which is very rare), Monte Carlo techniques can be used, provided that the cost of the mechanical analysis remains low. The use of meta-models allows us to reduce the cost of mechanical analyses by introducing an approximate analytical model.

In many real cases, calculating the reliability index is an effective means for analyzing industrial systems. A design point is obtained using a specific optimization algorithm, which can drive the mechanical model either directly or indirectly. After this step, probability can be estimated using the FORM/SORM techniques or using importance samplings in the neighborhood of the design point. In addition to reliability, this procedure allows designers access to the important mechanical and probabilistic parameters. This information is essential for optimizing the system, by taking into account the uncertainties at different levels of design, manufacturing, installation and maintenance.

Bibliography

[AFG 03] ASSOCIATION FRANÇAISE DE GENIE CIVIL, *Application des notions de fiabilité à la gestion des ouvrages existants*, CREMONA C. (ed.), Presses de l'Ecole Nationale des Ponts et Chaussées, Paris, France, 2003.

[AND 04] ANDRIEU-RENAUD C., SUDRET B., LEMAIRE M., "The PHI2 method; a way to compute time-variant reliability", *Reliability Engineering and System Safety*, vol. 84, no. 1, p. 75-86, 2004.

[BOÉ 09] BOÉRO J., SCHOEFS F., MELCHERS R., CAPRA B., "Statistical Analysis of Corrosion Process along French Coast", *ICOSSAR'09,* FURUTA H., FRANGOPOL D.M., SHINOZUKA M. (eds), Taylor & Francis, London, 2009.

[CEL 06] CELEUX G., MARIN J.-M., ROBERT C.P. "Iterated importance sampling in missing data problems", *Computational statistics & Data Analysis*, 50, 3386-3404, 2006.

[CEN 03] CEN, Eurocode 1, Actions on structures – Part 1–3: General actions – Snow loads, BS EN 1991-1-3, 2003.

[CEN 05] CEN, Eurocode 1, Actions on structures – Part 1–4: General actions – Wind actions, BS EN 1991-1-5, 2005.

[DIT 96] DITLEVSEN O., MADSEN H.O., *Structural Reliability Methods*, John Wiley & Sons, New York, USA, 1996.

[FAV 04] FAVRE J.L., 2004, *Sécurité des ouvrages – risques : modélisation de l'incertain, fiabilité, analyse des risqué*, Ellipses, Paris, France, 2004.

[HAG 91] HAGEN O., TVEDT L., "Vector process out-crossing as parallel system sensitivity measure", *J. Eng. Mech.*, vol. 117(10), p. 2201-2220, 1991.

[HAS 74] HASOFER A.M. and LIND N.C., "An exact and invariant first order reliability format", *J. Eng. Mech.*, ASCE, vol. 100, EM1, p. 111-121, 1974.

[LEM 09] LEMAIRE M., CHATEAUNEUF A. and MITTEAU J.C., *Structural Reliability*, ISTE Ltd, London and John Wiley & Sons, New York, 2009.

[LEM 00] LEMAIRE M., MOHAMED A., "Finite element and reliability: a happy marriage?", In *Reliability and optimization of structural systems, Keynote Lecture, 9th IFIP WG 7.5 Working Conference*, NOWAK A. and SZERSEN M. (eds), Ann Arbor, Michigan, USA, September 25-27, pp. 3-14, 2000.

[LEM 99] LEMING M.L., Probabilities of low-strength events in concrete, *ACI Str. J.*, 5-6/1999, pp. 369-377, 1999.

[MAR 09] MARACHE A., BREYSSE D., PIETTE C., THIERRY P., "Geotechnical modeling at the city scale using statistical and geostatistical tools: The Pessac case (France)", *Engineering Geology*, vol. 107, p. 67-76, 2009.

[MAT 08] MATTHIES H.G. "Structural damage and risk assessment and uncertainty quantification", *NATO-ARW Damage Assessment and Reconstruction After Natural Disasters and Previous Military Activities*, Sarajevo, Bosnia and Herzegovina, 5-9/10/2008, 2008.

[MEL 99] MELCHERS R.E. *Structural Reliability Analysis and Prediction*, John Wiley & Sons, New York, 1999.

[PEY 09] PEYRAS L., MERIAUX P. (coord). *Retenues d'altitudes*, Editions QUAE – Savoir Faire, Versailles, France, p. 309, 2009.

[SCH 08] SCHOEFS F., *Sensitivity Approach for Modelling the Environmental Loading of Marine Structures through a Matrix Response Surface*, Available online 19 June 2007, vol. 93(7), pp. 1004-1017, July 2008. doi: dx.doi.org/10.1016/j.ress.2007.05.006.

[SHI 91] SHINOZUKA M., DEODATIS G., "Simulation of Stochastic Processes by Spectral Representation", *Appl. Mech. Rev.*, vol. 44(4), April 1991.

[SMI 96] SMITH D., BIRKINSHAW M., HOBBS R., FISHER P., "ISO (draft) offshore design code: wave loading requirements. Discussion", *Transactions, Institute of Marine Engineers*, vol. 108(2), pp. 89-96, 1996.

[SUD 08a] SUDRET B., "Analytical derivation of the outcrossing rate in time-variant reliability problems", *Structure and Infrastructure Engineering*, vol. 4(5), p. 353-362, 2008.

[SUD 08b] SUDRET B., BERVEILLER M., "Stochastic finite element methods in geotechnical engineering", in *Reliability-Based Design in Geotechnical Engineering: Computations and Applications*, PHOON K.K. (ed.), Taylor & Francis, London, 2008.

[TAN 07] TANNERT C., ELVERS H.D., JANDRIG B., "The ethics of uncertainty. In the light of possible dangers, research becomes a moral duty", *EMBO reports,* vol. 8(10), p. 892-896, 2007.

PART 1

Qualitative Methods for Evaluating the
Reliability of Civil Engineering Structures

Introduction to Part 1

Structural reliability is a major concern for civil engineers, both during the design and construction of structures and throughout their working life. The way to gain a clear picture of this issue is through an understanding of a structure's operation and failure modes over time. Part 1 considers qualitative methods for evaluating the reliability of civil engineering structures. These methods can be used to analyze the operation of structures, their failure modes and failure scenarios that can lead to degraded function and to hazardous situations. These methods lead to a model of operational reliability for a structure, which consists of a representation of the modes of operation and malfunction of the system. These are qualitative concepts, and operational reliability models that serve as the basis for a quantitative analysis are built on the methods and tools developed in subsequent parts of this book.

Part 1 of this book is divided into three chapters. Chapter 1 discusses methods for system analysis and failure analysis, which can be used to obtain the failure modes. Chapter 2 presents the three main methods (the event tree, cause tree and bow-tie methods) that can be used to construct scenarios based around failure modes, followed by a presentation of methods for evaluating the criticality of a scenario. In an engineering context, these latter methods can be used to eliminate the least critical scenarios, enabling us to focus on analyzing those scenarios that pose the greatest risks. Chapter 3 gives an example application of the methods developed in the first two chapters, in the form of a hydraulic civil engineering installation.

Chapter 1

Methods for System Analysis and Failure Analysis

1.1. Introduction

System analysis methods can be used to develop an operational model for a system, known as a functional model, in order to describe its interactions with its environment. Building on this functional model, failure analysis methods aim to identify failure modes for a system, as well as their causes and consequences.

A range of failure analysis methods exist, such as preliminary hazard analysis [HAD 97], [MOR 02], [MOR 05], hazard and operability study [DES 06], [IMD 04], hazard analysis critical control point, or hazard identification [DES 03], etc. Later in this chapter, we will describe Failure Modes and Effects Analysis (FMEA), which is one of the most commonly-used of these techniques.

A system is defined in terms of the structural components (or their sub-systems) that it is constituted of and the functions that it performs, along with the temporal, spatial and functional boundaries used in the study. At each level of decomposition of the system (Figure 1.1), the structural components each perform functions that contribute to the global function of the system [CRE 03].

————————————

Chapter written by Daniel Boissier, Laurent Peyras and Aurélie Talon.

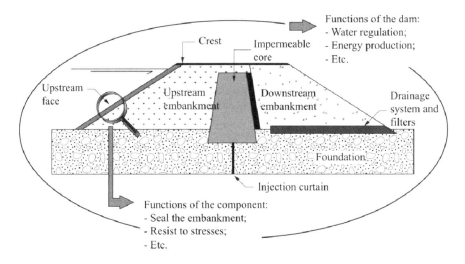

Figure 1.1. *Components and functions of a specific civil engineering structure:*
an example of a dam [PEY 03]

System analysis consists of two phases:

– structural analysis; and

– functional analysis.

NOTE: System analysis only has any meaning in the context of a particular concern or problem to be solved. The components and their inter-relationships cannot therefore be fully defined in a generic manner without reference to a specific problem.

In civil engineering, the systems under study are unique in terms of their geometry and their interactions with the environment, and so particular attention must be paid to structural analysis before beginning a functional analysis.

Structural analysis determines the most appropriate level of detail, or granularity, for the study that will enable the failure modes to be identified with an appropriate level of precision (the the most relevant to the goals of the user) during failure analysis.

The time taken by this study can be reduced if the system under study is geometrically similar to a system that has previously been studied.

1.2. Structural analysis

Structural analysis involves defining the bounds of a system, the scale of the study and the content of the system.

In the context of risk analysis, a system can be broken down into a set of sub-systems, an environment of causes and an environment of consequences (which may or may not be the same as the environment of causes, see Figure 1.2).

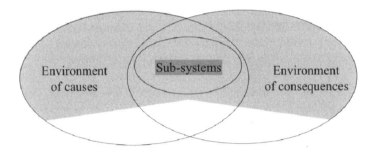

Figure 1.2. *System diagram in the context of risk analysis*

The shaded part of Figure 1.2 represents the system under study along with its boundaries. This shaded part only covers the set of sub-systems and environments that will be taken into account in the study.

It should also be noted that these three domains (sub-systems, environment of causes and environment of consequences) will evolve over time, since the sub-systems may be modified (repaired or maintained), the environment of causes may be modified (by climatic changes, or change of purpose of a subsystem), and the environment of consequences may also be modified (by construction of other products, change of standards or evolution in mentalities).

NOTE: The knowledge available on a system and the environment of causes and consequences may evolve over time.

1.2.1. *The sub-systems*

The sub-systems contain all the components (material, human, organizational and procedural) that can be acted upon (for example, a construction or an infrastructure for which the dimensions are controlled, inhabitants, skiers who can be directed using signage, or watercourses that can be controlled using hydraulic engineering works). Note that it is not necessarily possible to perform the same actions on every element that makes up the sub-system.

1.2.2. *Environments*

This describes all the elements of a given problem that are out of our control, such as rain, earthquakes, fire, conditions of use, etc.

The environment of causes includes all those components that have an effect on the sub-system. The environment of consequences includes all those elements on which the subsystem may have an effect. A given element, whether or not it belongs to a particular subsystem, may be both a cause and a consequence.

EXAMPLE 1.1. A sub-system could be an accommodation block. An element of the environment of causes could be the outbreak of a fire. The scenario of dangerous events could be the propagation of a fire that burns combustible materials within the accommodation and causes the release of toxic gases. The environment of consequences may include the occupants (asphyxiated as a result of this dangerous scenario) and the atmosphere (pollution by release of toxic gases and local heating due to the heat of the fire) and possibly even the fire-starter(s) themselves.

1.2.3. *Bounding the analysis*

In any analysis it is crucial to identify the boundaries of the system. Three different types of boundaries can be identified: spatial, temporal and engineering boundaries.

Spatial boundaries provide a geometric or geographical definition for the study. For example, we could consider the risk of collapse of an inhabited building. Here, the subsystem includes the foundations, the load-bearing structure and the roofing of the building. The spatial boundary of the subsystem could be the building envelope or, in another problem, it could be the plot of land, the city block, etc. Thus the environment of causes includes normal climatic conditions (precipitation, wind, snow, etc.), exceptional environmental conditions (earthquakes, fire, flood, etc.) and usage conditions (inhabitants, floors, etc.). The environment of consequences can be spatially restricted to the scope of the building and its users, and may or may not consider residents, cars, passers-by, groundwater, the ozone layer, etc.

Temporal boundaries define the time period over which the system is to be studied. This may be one or more years, one or more cycles, etc. For example, we are interested in the risk of collapse of a particular building during its service life, but not when it is being demolished.

The *engineering boundaries* reduce the scope of the investigation and analysis to match the abilities of the people performing the analysis and to the domain of definition of the decision variables. For example, in fire safety engineering there is little interest in evaluating the consequences of a building fire on the ozone layer; it is also pointless to propose a technical solution that cannot be implemented in practice.

1.2.4. *Scales of a study*

Five scales for a study can be identified:

– *functional scale*: this type of analysis aims to study the response of a system in terms of the purposes for which it was built. Depending on the problem in question, this analysis may include all the functions of a system (safety of goods and people), or only its primary functions (structural stability, acoustic insulation, etc.), or focus on just one particular function identified by the decision maker;

– *temporal scale*: the reliability of a system in fulfilling the functions for which it was or will be constructed depends on the time interval that is considered;

– *geometric scale*: a system analysis does not necessarily encompass the entire project; expertise or preliminary analysis may enable us to isolate within the project a specific geometric entity that is the key to our analysis;

– *phenomenological scale*: an undesirable scenario (series of events) is often the result of multiple causes, but more importantly it sets in motion a very wide range of natural phenomena: physical, behavioral, organizational, etc. These phenomena can be considered over a range of different scales (e.g. the initiation, bifurcation and propagation of a crack or length of a crack, or the amount of cracking within a structure);

– *logistic scale*: this analysis considers the availability, durability and maintainability of a system.

Once the boundaries and scales have been defined for the study, the content of the system(s) and the environment of causes and consequences must be determined. This phase involves listing all the components that make up each subsystem, defining their physical position and determining their interactions with other sub-systems or components of sub-systems.

EXAMPLE 1.2. Structural breakdown of an embankment dam with chimney drain into five components (Figure 1.3).

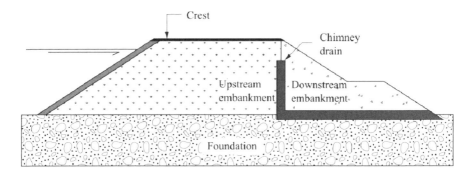

Figure 1.3. *Structural breakdown of an embankment dam with chimney drain [PEY 03]*

Once the interactions between components of a subsystem are known, the distribution of forces, temperature flows, etc., can be identified for the case of normal working and, consequently, the impact of the failure of a particular component on the other components of the subsystem can also be identified. An understanding of interactions between the subsystem and the environment of causes enables us to describe the effects of the environment on the subsystem, while an understanding of the interactions between the subsystem and the environment of consequences can be used to focus the study of the impact of failure of a subsystem on the components that make up this environment of consequences.

1.3. Functional analysis

1.3.1. *Principles of functional analysis*

Functional analysis helps us to understand the behavior of a system under study and to describe its behavior in a concise manner, and formally and exhaustively establishes the functional relationships within and outside a system. Functional analysis consists of two phases:

– external functional analysis; and

– internal functional analysis.

A wide range of methods of functional analysis exist that are well suited to industry, services or organizations, such as Reliasep, FAST, DEN, SADT, etc., but no specific method has been developed for civil engineering. An examination of various different functional analysis techniques reveals that these value analysis issues are widely applicable and are ideally suited to mechanical systems.

Of these, the APTE method (*Application aux Techniques d'Entreprise*, or *Application of Corporate Techniques*) has been used and adapted with success to civil engineering systems, particularly in applications involving the structural elements of a building [LAI 00] and hydraulic works [PEY 03], [SER 05].

1.3.2. *External functional analysis*

In external functional analysis, the work under study is treated as a "black box", and it is the interactions of this black box with its environment that are studied.

External functional analysis can be used to determine the functions performed by a system, which is treated in a global manner through the use of functional block diagrams.

The *principal functions* are the essential functions for which the system was designed: these are obtained by examining the environments interacting with the system.

EXAMPLE 1.3. The principal functions of a dam are to:

– store water within a reservoir;

– cope with flooding; and

– supply water to a manufacturing plant (or other external systems).

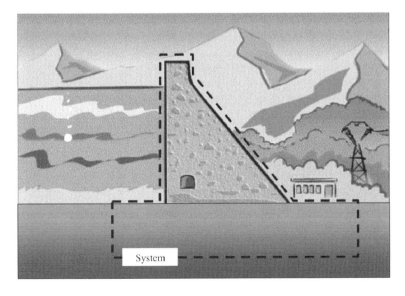

Figure 1.4. *Definition and limits of a gravity dam [PEY 03]*

Secondary functions result from the response of the system to externally imposed stresses, and are obtained by examining the interactions with the external environment.

EXAMPLE 1.4. The secondary functions of a dam are to withstand, evacuate, retain, drain, filter, etc.

The functional block diagram for the dam shown in Figure 1.4 is given in Figure 1.5.

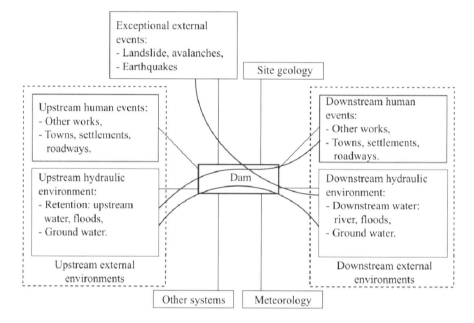

Figure 1.5. *Functional block diagram for a dam [PEY 03]*

1.3.3. *Internal functional analysis*

Following a global system analysis phase, internal functional analysis aims to identify the functions of each of its components. Flow analysis can be used to identify the secondary functions of the components.

Three types of flow can be identified:

– transport flows, which can be used to identify in functional terms the propagation of flows from the environment to the system components (for example, the functions of water-tightness or air-tightness for a window);

– load flows, which can be used to identify the functions of resistance to loads applied by the environment of causes or by connected components (for example, a dam's function of resistance to hydrostatic pressure);

– contact flows, which can be used to identify the functions that involve the mechanical and physiochemical properties of the components (for example, the function of resistance of steel damaged by corrosion caused by chloride ions in seawater).

Component	Functions
Impervious core	1 – Resistance to mechanical loads 1.1 – The impervious core resists the hydrostatic pressure transmitted by the upstream embankment 1.2 – The impervious core resists the under-pressure from the upstream foundations 1.3 – The impervious core resists the force of the upstream embankment 1.4 – The impervious core resists the force of the downstream embankment 1.5 – The impervious core resists the forces transmitted by the crest 2 – Limit hydraulic flows 2.1 – The impervious core limits infiltration from the upstream embankment 2.2 – The impervious core limits infiltration from the upstream foundation 2.3 – The impervious core limits infiltration of rainwater from the crest

Table 1.1. *Functions of the impervious core of an embankment dam [PEY 03]*

These various flows can be represented in the form of functional block diagrams; an example of this is given in Figure 1.6 for an embankment dam.

At the end of functional analysis, all the functions achieved by the components of the subsystem(s) have been determined. These are combined to give functional analysis tables; an example of this is shown in Table 1.1 for an embankment dam.

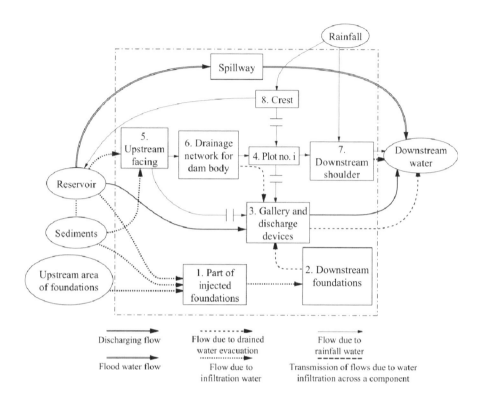

Figure 1.6. *Functional block diagram for hydraulic flow transport in an embankment dam with upstream facing [PEY 03]*

1.4. Failure Modes and Effects Analysis (FMEA)

1.4.1. *Principles of FMEA*

Failure Modes and Effects Analysis (FMEA) has a range of applications in civil engineering. Particularly worthy of mention are its uses in hydraulic engineering works [PEY 06], [PEY 03], [SER 05] and building components [LAI 00], [TAL 06a], [TAL 07].

It has attracted the interest of several national and international research centers in the field of civil engineering. The state of the art for these approaches was the subject of CIB report W80 "Service Life Methodologies – Prediction of Service Life for Buildings and Components" [TAL 06b], with the aim of encouraging the application of this method in the construction industry.

FMEA, as it has been developed and standardized in industry [DEP 80], [IEC 85], [AFN 86], is a method of inductive analysis for potential system failures. It systematically considers each {function; component} pair for a system, one after the other, along with analysis of its failure modes.

The results of FMEA analysis are presented in tabular form, specially tailored to the system of system under study, and their format can vary considerably from one field to another [FAU 04], [ISD 90].

In industry, FMEA studies fit in with a production objective, where the number one aim is to eliminate failures associated with each stage of manufacturing, e.g. design and implementation errors, and deviations from the standards and rules of hygiene and safety.

In civil engineering, it has been necessary to adapt FMEA in order for it to be applied to the field of structures and buildings. The FMEA approach used in civil engineering consists of two phases, which are summarized in Figure 1.7.

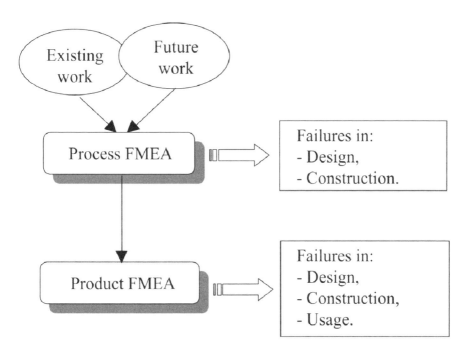

Figure 1.7. *FMEA approach applied to civil engineering*

Process FMEA considers a structure after construction has been completed, and looks for potential failure modes associated with design and construction defects.

Product FMEA then considers a structure whose quality may vary depending on how it was built and reveals, in addition to failures associated with use of the structure, the influence of the design and construction process on its future behavior, up until the point at which it may be demolished.

1.4.2. *Process FMEA*

Process FMEA involves determining, with the help of a table, the main operations of the design and construction process for each component of the structure, the failure modes of these operations and the effects of these failure modes.

EXAMPLE 1.5. Process FMEA applied to embankment dams.

Process FMEA results in a description of the structure once its construction has been completed. In particular, it can be used to determine the potential failure modes associated with design and construction failures that would become the causes of failures during the usage phase.

Component	Process operation	Failure mode	Failure effects
Grout curtain	Design: – geological surveying – geotechnical study – study of reinforcement (depth, density, grout, etc.). Construction: – drilling operations – grouting operations.	Design: – insufficient surveying – limited geotechnical study – inadequate reinforcement study. Construction: – insufficient density – insufficient depth – inappropriate grout quality – inappropriate grouting pressure, – insufficient lateral extension.	– insufficiently watertight – lateral bypassing – bypassing underneath – disintegration of grout curtain – cracking of grout curtain.

Table 1.2. *Process FMEA (extract for the grout curtain of an embankment dam) [PEY 03]*

1.4.3. *Product FMEA*

The aim of Product FMEA is to identify the various failure modes of each subsystem when it is in service. This is connected to the design and construction process, but also to hazards that may occur during its use.

The results are presented in the form of a table showing:

– components of the system;

– functions of each component, obtained through functional analysis;

– failure modes for each component, corresponding to failures or degradation of a function;

– possible causes of failure, determined using functional block diagrams, and originating from the various flows (loads, hydraulic, hydrodynamic and thermal), contact with external environment or neighboring components, the intrinsic state of the component or the design or construction process for the component;

– possible consequences of the failure, determined in a deductive manner.

In summary, Process and Product FMEA reveal the failure modes of a system, their causes, their consequences, and potentially series of failure modes.

In practice, two formats for the Product FMEA tables are used, with the first being more useful for the design phase and the other more useful for the usage phase:

– "function" analysis: the aim here is to identify the failure modes leading to reduced performance of functions of the system under study; thus the table consists of columns for "function", "component", "stage", "mode", "cause" and "effect"; an example of this is shown in Table 1.3;

– "component" analysis: the aim here is to identify the failure modes that affect each component of the system under study; thus the table consists of columns for "component", "function", "mode", "cause" and "effect"; this format has been used for hydraulic works.

EXAMPLE 1.6. FMEA for a concrete wall.

This example illustrates the application of FMEA to a concrete wall, performed at the granularity of the wall over its service lifetime, and including causes having their origin in the construction phase. The function of the wall that we will consider here is its mechanical resistance. Several different failure modes can be imagined for this: segregation, defect of coating, carbonation, cracking, etc.

The aim is to identify causes of failure modes. These may have their origins within the design/construction phase, identified using Process FMEA (concrete construction), in the environment of causes for the system (humidity, rainfall), or in the intrinsic state of the system (corrosion, salt swelling, concrete spalling).

The effects of failure modes involve a loss of mechanical resistance: cracking, irreversible movements, etc., and they may in the most severe cases lead to a loss of stability.

Function	Component	Stage	Failure mode	Cause	Effect
Mechanical resistance	Structure	0	Segregation	Construction	Reduction in mechanical resistance
	Structure	0	Insufficient cladding	Construction	
	Reinforcement of external skin	1	Corrosion	Carbonation of the structure (etc) Stage 1	
	Structure	2	Cracking	Corrosion of the reinforcement of external skin (etc) Stage 2	
	Interface between structure and reinforcement of external skin	3	Alkali reaction	Structural cracking (etc) Stage 3	
	Structure	4	Spalling	Formation of swelling salts (etc) Stage 4	
	Structure	5	Pieces falling off	Structural spalling (etc) Stage 5	
	Structure	1	Lixiviation	Rainfall, pure water	
	Insulation	1	Crushing	Hard shock (high intensity)	
	Lining	2	Crushing	Crushing of insulation (etc) Stage 1	
	Adhesive layer	3	Crushing	Crushing of lining (etc) Stage 2	

Table 1.3. *Extract from the FMEA for a concrete wall [TAL 06a]*

The column entitled "Stage" in Table 1.3 makes it easier to identify a series of failure modes. For example, the effect of "corrosion of reinforcement of the external skin" in Stage 2 leads to the effect of "structural cracking" in Stage 3. Stage 0 corresponds to effects that take place during the construction phase of the concrete wall, while Stages 1 to 6 correspond to effects taking place during the service life of the wall.

The main advantages and limits of FMEA are as follows:

– the exhaustive nature of the results (failure modes, causes, effects): this is the main advantage of the method, which involves a systematic approach to the identification of failure modes; it represents the current understanding, at a given moment, of the functioning and failure modes of a system;

– the corollary to this advantage is that it is entirely built on the structural and functional understanding of the system under study, and hence also on the quality of the preliminary system analysis that has been performed; it is therefore necessary to take particular care during the system analysis phase;

– the FMEA approach can be used to develop a basis of information to define, improve, correct and validate a design, process or method right through from its design to the end of its service life;

– combined or dynamic effects are difficult to take into account; for this reason, FMEA is generally followed by a method that enables failure scenarios to be modeled.

1.5. Bibliography

[AFN 86] AFNOR, Techniques d'analyse de la fiabilité des systèmes – Procédures d'analyse des modes de défaillance et de leurs effets (AMDE), NF X60-510, 1986.

[DEP 80] DEPARTMENT OF DEFENSE, Military standard procedures for performing a failure mode, effects and criticality analysis, MIL-STD-1629A, 1980.

[DES 03] DESROCHES A., LEROY A., VALLEE F., *La gestion des risques: principes et pratiques*, Hermès, Paris, France, 2003.

[DES 06] DESROCHES A., LEROY A., QUARANTA J.-F., VALLEE F., *Dictionnaire d'analyse et de gestion des risques*, Hermès, Paris, France, 2006.

[FAU 04] FAUCHER J., *Pratique de l'AMDEC*, Dunod, Paris, France, 2004.

[HAD 97] HADJ-MABROUK H., *L'analyse préliminaire de risques*, Hermès, Paris, France, 1997.

[IEC 85] IEC, *Techniques d'analyse de la fiabilité des systèmes – procédure d'analyse des modes de défaillance et de leurs effets (AMDE)*, IEC 60812, 1985.

[IMD 04] IMᴅR-SᴅF, Groupe de travail Management Méthodes Outils Standard (M2OS) – Fiches Méthodes [online], IMdR-SdF, 2004. Available at: http://www.imdr-sdf.asso.fr/v2/extranet/images/Fiches_Methodes-b.pdf (accessed 27/04/2009).

[ISD 90] ISᴅF. *AMDEC – Analyse des modes de défaillances, de leurs effets et de leur criticité – Pedagogical guide*. Institut de la sûreté de fonctionnement (ISdF), Nanterre, France, 1990.

[LAI 00] Lᴀɪʀ J., Evaluation de la durabilité des systèmes constructifs du bâtiment, Civil Engineering Thesis, Blaise Pascal University – Clermont II, CSTB & LERMES, Clermont-Ferrand, France, 2000.

[MOR 02] Mᴏʀᴛᴜʀᴇᴜx Y., "Analyse préliminaire de risque", *Techniques de l'ingénieur*, no. SE 4010, 2002.

[MOR 05] Mᴏʀᴛᴜʀᴇᴜx Y., "La sûreté de fonctionnement: méthodes pour maîtriser les risques", *Techniques de l'Ingénieur*, no. BM 5008, Paris, France, 2005.

[PEY 03] Pᴇʏʀᴀs L., Diagnostic et analyse de risques liés au vieillissement des barrages – Développement de méthodes d'aide à l'expertise, Civil Engineering Thesis, Blaise Pascal University – Clermont II, LERMES & CEMAGREF, Clermont-Ferrand, France, 2003.

[PEY 06] Pᴇʏʀᴀs L., Rᴏʏᴇᴛ P., Bᴏɪssɪᴇʀ D., "Dam ageing diagnosis and risk analysis: development of methods to support expert judgement", *Canadian Geotechnical Journal*, vol. 43, p. 169-186, 2006.

[SER 05] Sᴇʀʀᴇ D., Evaluation de la performance des digues de protection contre les inondations. Modélisation de critères de décision dans un système d'information géographique, Geographic Information Science Thesis, Marne-la-Vallée University, Paris, France, 2005.

[TAL 06a] Tᴀʟᴏɴ A., Evaluation des scénarios de dégradation des produits de construction, Civil Engineering Thesis, Blaise Pascal-Clermont II University, CSTB & LGC, Clermont-Ferrand, France, France, 2006.

[TAL 06b] Tᴀʟᴏɴ A., Cʜᴇᴠᴀʟɪᴇʀ J.-L., Hᴀɴs J., "Failure modes effects and criticality analysis research for and application to the building domain", publication CIB no. 310, Working commission 80 "Service life methodologies – Prediction of service life for buildings and components", *International Council for Building Research Studies and Documentation*, Rotterdam, The Netherlands, 2006.

[TAL 07] Tᴀʟᴏɴ A., Bᴏɪssɪᴇʀ D., Lᴀɪʀ J., "Service life assessment of building components: application of evidence theory", *Canadian Journal of Civil Engineering*, vol. 35(3), p. 66–70, 2007.

Chapter 2

Methods for Modeling Failure Scenarios

2.1. Introduction

Once the failure modes of a system have been identified, methods for modeling failure scenarios can be used to build series of failure modes – failure scenarios – that may lead to global system failure. Of the various methods for modeling failure scenarios (such as the reliability diagram method, the failure combination method, the cause-effect diagram method, etc.) this chapter will only consider the three main methods that are most widely used in industry and civil engineering:

– the *event tree method*, which can be used to describe the scenarios that ensue from an initial trigger event;

– the *fault tree method*, which gives a detailed description of the scenario leading up to an unwanted event;

– the *bow-tie method*, which describes the scenarios leading to an unwanted event (fault tree) and the series of effects resulting from that same event (event tree).

The use of these three methods in the field of civil engineering is analogous to the way they are employed in industry, and no specific adaptations are required in order to apply them to civil engineering.

Chapter written by Daniel BOISSIER, Laurent PEYRAS and Aurélie TALON.

2.2. Event tree method

The event tree method is an inductive method: it starts with an event at the top of the tree, and consists of all the combinations of events that are triggered by the occurrence of this initial event. It is often used to determine all the consequences of an unwanted event.

The event tree method finds applications in the operational reliability of systems whose functioning is roughly binary (either "running" or "broken") or discrete (the events or changes occur at specific dates in time), and submitted to time evolution. This method is described in a wide range of texts, such as [MOR 05], [IMD 04], [DES 03], [MOR 02], [MOR 01], [ZWI 96], [MOD 93], [VIL 88].

Event trees are conventionally constructed horizontally, from left to right, starting from an initiating event. The tree is built up chronologically in an inductive manner, by studying the behavior (functioning or malfunction) of each system component. The functioning or malfunction of a component thus corresponds to an event, and a scenario is formed of a combination of several events. Event trees can be used to determine the sequence of events leading up to a final event.

The quality of the operational reliability model obtained for a system, based on an event tree, depends on the quality of the preliminary functional analysis and failure analysis. In order to ensure that the failure analysis is exhaustive, it is recommended that a FMEA be performed before following the event tree method. Prior application of FMEA has the advantage that the event tree is built up naturally from sequences of failure modes obtained through FMEA.

In the course of a quantitative risk analysis, the probability of each event in the tree occurring is evaluated, and the probability of a scenario occurring is then equal to the product of the probabilities that each individual event making up the scenario will occur[1]. The event tree method can be used to explore how a scenario unfolds, and to evaluate the effect of the introduction of barriers[2] on the frequency of appearance of the scenario and on its consequences.

As an illustration, Figure 2.1 shows an event tree corresponding to the effects of an initiating event (fire door left propped open) on a system (fire door and user).

On the topmost branch of this tree, all the barriers that are in place (fire alarm, fire detection, etc.) function correctly and, as a result, nobody will be injured.

_____

1 Assuming that each event is independent in probabilty terms.
2 A barrier is a component whose aim is to reduce or eliminate a risk.

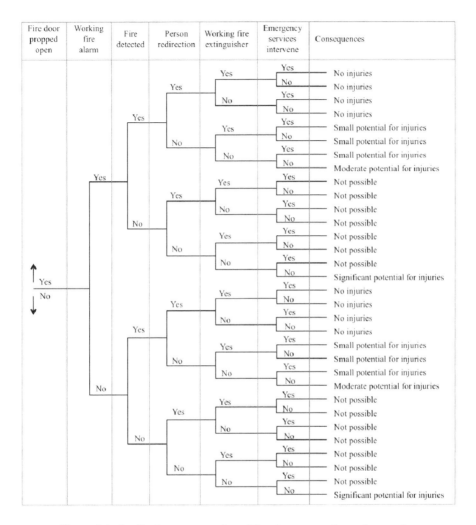

Figure 2.1. *Qualitative event tree for a "fire door propped open" scenario, for a "fire door and user" system*

Conversely, a failure (lowermost branch of the tree) of every safety barrier may lead to a significant number of injuries. The set of possible failure/non-failure combinations for the safety barriers is shown between these uppermost and lowermost branches.

This is a very easy method to employ, being based on an intuitive approach. It enables the building of failure scenarios from chained sequences of events. The quality of the analysis depends on the quality and exhaustivity of events that are considered, representing the interactions between the environment and the system, and the potential behavior of the system.

This analysis, preferably a group analysis, then provides information to lead a preliminary system quality analysis and a failure analysis method such as FMEA.

2.3. Fault tree method

The fault tree method (or cause tree method) is a deductive method: it starts with an event lying at the top of the tree, and determines all the combinations of causes that could explain the occurrence of this event. It is often used to determine all the causes that can lead to an unwanted event.

Drawing up a fault tree has long been considered an art practiced by an analyst, and as such it can be difficult to ensure that all failure modes are considered exhaustively. This method is widely described; see in particular [GIR 06], [MOR 05], [MOR 01], [MOR 02], [VIL 88], [ZWI 99].

The fault tree approach includes an information acquisition phase and a tree construction phase. In what follows we will only consider fault trees associated with unwanted events.

2.3.1. *Information acquisition*

Failure causes that contribute to the occurrence of a failure in the fault tree method are generally determined through practice feedback. Causes can also be recognized directly through discussions and interviews with experts in the field.

Use of the FMEA method is an alternative, complementary approach for obtaining information.

It is also recommended that a complete system analysis be conducted before applying the fault tree method.

2.3.2. *Fault tree construction*

By convention, a fault tree is presented vertically, with the final failure under study lying at the top of the tree. The first phase of fault tree construction therefore involves determining the eventual failure modes that it is hoped will be avoided.

The second phase involves tracing back, step by step, to the originating causes. At each phase of construction, the aim is to answer the question "what must have happened in order that...?" Other questions that must then be answered are "are there any other causes?" and "could the failure we are interested in have occurred if one of the causes had not been present?"

Construction of the tree is stopped when original causes are reached that are not an inherent part of the system under study (external circumstances). In practice, the analysis is stopped when the causes found can be eliminated or their probability reduced through industrial or technological solutions.

EXAMPLE 2.1. Metal facing screwed to wooden beams.

This example uses the fault tree method to study the fall of a piece of metal facing fixed to wooden beams, which can be explained by causes associated with exceptional external stresses, unanticipated aging of the system or a construction fault (Figure 2.2).

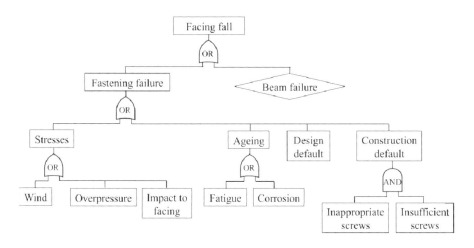

Figure 2.2. *Example fault tree for a metal facing screwed to wooden wall plates*

The main points of interest of this method are:

– it is easy and intuitive to implement;

– it favors use of expertise to identify causes; and

– it favors investigation and evaluation of prevention or protection measures intended to avoid the appearance of the unwanted outcome.

The main limitations of this method are that:

– investigation of causes does not occur in a systematic manner, and does not therefore guarantee exhaustive consideration of failure modes. It must therefore draw on expertise and a well-documented experience feedback;

– it does not offer a chronological representation of events;

– it is a binary method, in other words it involves events that either do or do not occur; there is no intermediate state.

2.4. Bow-tie method

The bow-tie method describes accident scenarios that may unfold, starting from the initial causes of an accident and tracing them forwards to their consequences, according to the identified goals. The "bow-tie" is a tree-type approach that combines a fault tree and an event tree (Figure 2.3).

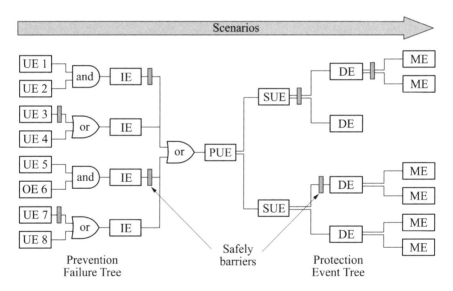

UE: Unwanted Event; OE: Ongoing Event; IE: Initiating Event;
PUE: Primary Unwanted Event; SUE: Secondary Unwanted Event;
DE: Dangerous effect; ME: Major effect

Figure 2.3. *Scenario representation using a bow-tie diagram [COM 08]*

This method is described in a range of publications, in particular [PRE 09], [COM 08], [COU 04], [INE 03]. The central point in the bow-tie diagram is the primary unwanted event (PUE). The left-hand part of the bow-tie then takes the form of a failure tree, whose aim is to identify the causes (initiating events – IE) of this primary unwanted event. The right-hand part of the bow-tie investigates the consequences of this primary unwanted event, and potentially secondary unwanted events (SUE), using an event tree. On this diagram, safety barriers are shown in the form of vertical lines in order to symbolize the fact that they guard against the development of an accident scenario. In this representation, each route leading from an originating failure (undesirable or ongoing) to the appearance of visible problems (major effects), represents a specific scenario associated with the primary unwanted event under consideration.

The bow-tie, which makes direct use of failure and event trees, should be built up according to the same guidelines. Since it is a relatively cumbersome tool to employ, its use is generally restricted to events that are considered particularly critical, for which a high level of understanding of the risks is required.

The bow-tie method is employed following an initial study of failures and primary unwanted events, carried out using methods such as preliminary risk analysis or FMEA. The principles for implementing the bow-tie method (Figure 2.4) follow below.

Stage 1 involves a system analysis followed by failure mode analysis, which might for example be performed using FMEA.

Stage 2 uses failure tree and event tree methods (defined earlier).

Stage 3 involves a criticality analysis, described in section 2.5. This analysis draws on practice feedback and expertise.

Stage 4 involves identifying prevention and protection barriers. *Prevention barriers* are intended to reduce the probability of failure causes occurring. For example, preventative avalanche initiation (artificial triggering) and ski run signage are preventative barriers against the dangers of an avalanche. *Protection barriers* are intended to limit the consequences of a failure. For example, the installation of snow nets and construction of an avalanche protection gallery are protective barriers against the effects of an avalanche.

Stage 5 involves re-evaluating the system to take account of the effects of preventative and protective barriers, and to obtain a new estimate for the probabilities of failures occurring, along with their severities. The effectiveness of the barriers is, however, variable.

Depending on the result of Stage 5, **Stage 6** can be used to confirm or reject the preventative and protective barriers that have been proposed. If the criticality of the failures is considered acceptable, then the process moves on to Stage 7, otherwise new barriers must be introduced (which involves returning to Stage 4).

Stage 7 generates a summary of the risk analysis as performed using the bow-tie method, for example by writing a risk prevention plan.

The bow-tie method can be illustrated with a simplified example of the use of an embankment dam:

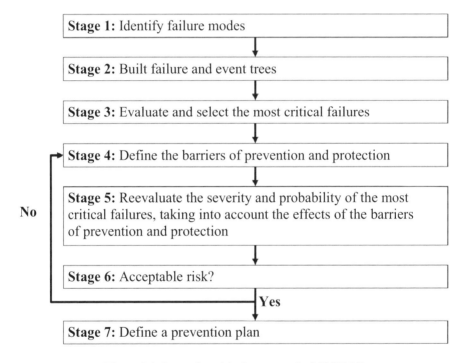

Figure 2.4. *Principles of the bow-tie method [COU 04]*

Stage 1: the main mechanisms for dam failure during use are internal erosion, external erosion and sliding collapse. The unwanted effects are:

– flooding of downstream elements; and

– degradation that prevents use of the retained water supply.

Stage 2: see sections 2.2 and 2.3 – the failure and event trees are not repeated here.

Stage 3: the three failure mechanisms are considered "extremely serious" but to have a low probability of occurring. For this illustration we will only consider the failure mechanism involving internal erosion.

Stage 4: the risks associated with internal erosion can be reduced by:

– monitoring the drainage rate;

– monitoring the piezometry; and

– monitoring the water level behind the dam.

Barriers that can limit the risk level may include:

– the drainage rate, which must be kept below a predetermined limit;

– the interstitial pressure, which must not exceed a threshold level for the pressure cells and pressure meters built into the embankment of the dam; and

– the floodwater level, which must not be higher than the maximum water level.

Stage 5: as long as these thresholds are respected, it can be assumed that the probability of failure will remain low, as long as the interval between two inspections is shorter than the interval between the thresholds being exceeded and the occurrence of the unwanted event. Conversely, when these thresholds are exceeded, a state of unacceptable risk develops.

Stage 6: under these conditions, the risk is now acceptable.

The bow-tie method offers a concrete visualization of accident scenarios that may unfold, starting from the initial causes of the accident and following them through to consequences in terms of the identified roles of the structure. Because of this, this tool clearly highlights the operation of safety barriers that guard against such accident scenarios, and can be used to demonstrate that the risks have been managed appropriately.

2.5. Criticality evaluation methods

In its original form, criticality involves quantifying the consequences of failure modes identified through FMECA (Failure Modes, Effects and Criticality Analysis). The concept of criticality can be generalized to failure scenarios for a system: it involves measuring the risk associated with the occurrence of a particular scenario.

This section summarizes the various formulations developed to evaluate the criticality of a particular scenario, along with practical considerations relevant to civil engineering applications.

2.5.1. *Criticality formulation*

The criteria used to evaluate criticality depend on the systems under study. Formulations involving two and three criteria are often used, and these will be described below.

2.5.1.1. *Two-criteria formulation*

The criticality formulation introduced in [SCT 03] and [FAU 04], and by Farer in the 1960s, takes into account the frequency of occurrence of a failure, and the severity of its consequences. These two criteria are evaluated using scoring grids, which may be qualiative or quantitative (Tables 2.1 and 2.2). The criticality is then determined using a grid that combines the values of these two criteria (Figure 2.5). The two scoring grids in Tables 2.1 and 2.2 were developed to estimate the frequency of occurrence and severity of consequences for failure modes of double glazing (identified using FMEA). Estimates for the frequency of occurrence and their severity are based on information obtained from double glazing manufacturers. Depending on the available information, the scoring grids may be quantitative (Table 2.1) or qualitative (Table 2.2).

Occurrence	Frequency of failure
0	Physically impossible
1	1 in 1 million
3	1 in 100,000
5	1 in 10,000
7	1 in 1,000
8	1 in 100
9	1 in 10
10	1 in 2

Table 2.1. *Scoring grid for occurrence rate – double glazing [HAG 02]*

Severity	Severity of consequences for failure modes
1	Effects present but not visible
3	Minor inconvenience for the client
5	Significant inconvenience for the client
9	Injury
10	Serious injury

Table 2.2. *Scoring grid for severity – double glazing [HAG 02]*

Several different scales are in common use, from 0 to 100, from 0 to 10 and from 0 to 4; the choice of scale must depend on the detail available in the estimation that is desired or feasible to obtain.

Figure 2.5 shows a three-level criticality grid (pale gray, darker gray and very dark gray) obtained by combining the frequency, estimated on a six-level scale, and the severity, also estimated on a six-level scale. This criticality grid has the advantage of being progressive, in contrast to a binary approach where something is "acceptable or unacceptable". For example, according to a maintenance budget distribution policy, those scenarios whose criticality is marked in very dark gray should be dealt with within a year, those scenarios in medium gray should be dealt with in the next three years, and those scenarios in light gray need not be dealt with.

2.5.1.2. *Three-criteria formulation*

The three criteria considered here are the frequency/occurrence of failures, their severity and their detectability or otherwise. There are then various approaches for evaluating and combining these criteria.

2.5.1.2.1. First approach [PEL 97], [LAS 01], [FAU 04], [GAB 09]

The most commonly used approach involves evaluating each criterion (O – Occurrence, S – Severity, D – Detectability) independently with the help of qualitative or quantitative scoring grids. The criticality (C) is defined as the product of these three criteria:

$$C = O \times S \times D \qquad\qquad [2.1]$$

Each failure is then classified in a binary manner using a criticality threshold, written $C_{thresh,}$ that is selected by the analyst:

– *if* $C < C_{thresh,}$ then the failure is considered acceptable;

– *otherwise* the failure is considered unacceptable (critical failure).

2.5.1.2.2. Second approach [BOW 95], [PIL 03], [GUI 04]

The values for the three criteria of criticality are evaluated with the help of two-level scoring grids. Thus, the occurrence, severity and undetectability of each failure mode are quantified on a scale of 1 to 10. These three values are then transformed into five qualitative classes (minor, weak, moderate, high and very high) in the case of occurrence and severity, and six classes (very high, high, moderate, weak, very weak and undetectable) in the case of undetectability. The confidence function associated with detectability is shown in Figure 2.6 as an example.

Frequency							
1. Failure very frequent	16	26	36	46	56	66	
2. Failure fairly frequent	15	25	35	45	55	65	
3. Failure infrequent	14	24	34	44	54	64	
4. Failure very infrequent	13	23	33	43	53	63	
5. Failure rare	12	22	32	42	52	62	
6. Failure extremely rare	11	21	31	41	51	61	
	1. Negligible effect	2. Minor effect	3. Significant effect	4. Serious effect	5. Major effect	6. Catastrophic effect	**Severity**

Figure 2.5. *Example criticality matrix – ammonia refrigeration plane [PEL 97]*

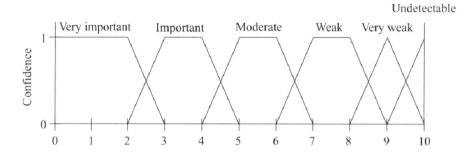

Figure 2.6. *Mapping function for detectability, taken from [BOW 95]*

Criticality is defined on a scale of ten values, in terms of six qualitative classes (unimportant, minor, weak, moderate, high and very high); its value is obtained either:

– using predefined rules determining the classes to which each of these three criteria belong (Table 2.3); for example, if the occurrence is weak, the severity is very high and the detectability is low, then the risk and hence the criticality are both high; or

– using the maximum likelihood method [BOW 95] applied to fuzzy subsets.

Rule	Occurrence	Severity	Detectability	Risk
1	Low	Moderate	Undetectable → Moderate	Moderate
2	Low	High	Undetectable → Moderate	High
3	Low	Very high	Undetectable → Low	Very high
4	Low	Very high	Moderate	High
5	Moderate	Moderate	Low → Moderate	Moderate
6	Moderate	High	Undetectable → Weak	Very high
7	Moderate	High	Moderate → Very high	High
8	Moderate	Very high	Undetectable → Moderate	Very high

Table 2.3. *Rules for determining criticality [BOW 95]*

2.5.1.2.3. Third approach [TAL 06]

Detectability may be integrated into the severity criterion and the duration of a scenario may be relevant in determining its criticality. Thus, this third approach assumes that the criticality depends on three criteria: the *probability of occurrence* of a scenario, its *duration* and its *severity*.

This approach was introduced in the construction domain [TAL 06]. The criteria of probability of occurrence and duration of the scenario may either be estimated with the help of scoring grids, or evaluated by data fusion (see Part 2).

The severity of a scenario is evaluated using six criteria which correspond to six essential requirements of the construction products directive [CST 93]. These criteria are evaluated using scoring grids. Criticality evaluation thus involves:

– a brief evaluation where the scenarios are classified into three categories of criticality (major, significant and minor) using criticality thresholds determined by the analyst; and

– a deeper evaluation where scenarios are prioritized in terms of their criticality, determined by the weighted product of the probability, duration of the scenario and the severity of its consequences.

2.5.2. *Civil engineering considerations*

The most widely used methods of evaluating criticality in the field of civil engineering are criticality grids (Figure 2.5) and the product of the occurrence, severity and detectability. These criteria are evaluated using scoring grids, tailored to the specific application. Prioritization of scenarios is determined either using thresholds on the criticality grids or based on the product of the three criteria (occurrence, severity and detectability, or probability, duration and severity).

Examples of criticality evaluation, using criticality grids, can be found in:

– [PEY 03] for dams;

– [CHO 07] for fire risk on construction sites.

Examples of criticality evaluation, using the product of three criteria, can be found in [LAI 00], [HAG 02], [LAI 02a], [LAI 02b], [LAI 03a], [LAI 03b], [LAY 98], [TAL 04], [TAL 06] for sealed systems, wooden and PVC windows, double glazing, metal cladding, solar collectors, brick and concrete walls.

2.6. Bibliography

[BOW 95] BOWLES J.B., PELAEZ C.E., "Fuzzy logic prioritization of failures in a system failure mode, effects and criticality analysis", *Reliability Engineering and System Safety*, vol. 50, p. 203–213, 1995.

[CHO 07] CHORIER J., Diagnostic et évaluation des risques incendie d'une construction et de sa mise en sécurité, Civil Engineering & Habitat Science Thesis, Savoy University, Chambéry, France, 2007.

[COM 08] COMTE G., "Méthodologie d'analyse de risques" [online], 2008, available at: http://www.qse-france.com (accessed 27/04/2009).

[COU 04] COURONNEAU J.-C., "Mise en œuvre de la nouvelle approche d'analyse des risques dans des installations classées (Principes généraux pour l'élaboration des études de dangers)" [online], 2004, available at: http://www.fluidyn.com (accessed 27/04/2009).

[CST 93] CSTB, "Directive européenne sur les produits de construction et documents interprétatifs", *Cahiers du CSTB*, no. 2704, 1993.

[DES 03] DESROCHES A., LEROY A., VALLEE F., *La gestion des risques: principes et pratiques*, Lavoisier, Paris, France, 2003.

[FAU 04] FAUCHER J., *Pratique de l'AMDEC*, Dunod, Paris, France, 2004.

[GAB 09] GABRIEL M., "AMDEC" [online], 2009, available at: www.cyber.uhp-nancy.fr/demos/MAIN-003/chap_deux/index.html (accessed 27/04/2009).

[GIR 06] GIRAUD M., "Sûreté de fonctionnement des systèmes – Analyse des systèmes non réparables", *Techniques de l'Ingénieur*, no. E 3851, Paris, France, 2006.

[GUI 04] GUIMARAES A.C.F., LAPA C.M.F., "Effects analysis fuzzy inference system in nuclear problems using approximate reasoning", *Annals of Nuclear Energy*, vol. 31, p. 107–115, 2004.

[HAG 02] HAGE R., "Capturing insulating glass failure events using failure modes and effects analysis and event trees", *IGMA Technical Division Meeting*, 2002.

[IMD 04] IMdR-SdF, "Groupe de travail Management Méthodes Outils Standard (M2OS) – Fiches Méthodes" [online], IMdR-SdF, 2004, available at: http://www.imdr-sdf.asso.fr/v2/extranet/images/Fiches_Methodes-b.pdf (accessed 27/04/2009).

[INE 03] INERIS, "Outils d'analyse des risques générés par une installation industrielle" [online], INERIS, 2003, available at: http://www.ineris.fr/ (accessed 27/04/2009).

[LAI 00] LAIR J., Evaluation de la durabilité des systèmes constructifs du bâtiment, Civil Engineering Thesis, Blaise Pascal University – Clermont II, CSTB & LERMES, Clermont-Ferrand, France, 2000.

[LAI 02a] LAIR J., CHEVALIER J.-L., "Failure Mode Effect and Criticality Analysis for Risk Analysis (Design) and Maintenance Planning (Exploitation)", *9th Durability of Building Materials and Components (9DBMC)*, Brisbane, Australia, 17–21 March 2002.

[LAI 02b] LAIR J., "Failure Modes and Effects Analysis for Electrochromic and Gaso-chromic Glazings", *SWIFT project*, WP2 Task 3: Reliability assessment, 2002.

[LAI 03a] LAIR J., Failure modes and effect analysis, service life prediction, report D4-C2-jl-01 Draft 2, IEA Task 27, Project C2: Failure Mode Analysis, 2003.

[LAI 03b] LAIR J., TALON A., LAPORTHE S., "Fiabilité et Durabilité des Systèmes Solaires Thermiques", *Action 3 de la convention ADEME*, no. 01.05.128.

[LAS 01] LASNIER G., *Gestion industrielle et performances*, Hermès, Paris, France, 2001.

[LAY 98] LAYZELL J., LEDBETTER S., "FMEA applied to cladding systems – Reducing the risk of failure", *Building Research and Information*, no. 26(6), p. 351–357, 1998.

[MOD 93] MODARRES M., *What Every Engineer Should Know About Reliability and Risk Analysis*, Marcel Dekker, New York, USA, 1993.

[MOR 01] MORTUREUX Y., "La sûreté de fonctionnement: méthodes pour maîtriser les risques", *Techniques de l'ingénieur*, no. AG 4670, Paris, France, 2001.

[MOR 02] MORTUREUX Y., "Arbres de défaillance, des causes et d'événement", *Techniques de l'ingénieur*, no. SE 4050, Paris, France, 2002.

[MOR 05] MORTUREUX Y., "La sûreté de fonctionnement: méthodes pour maîtriser les risques", *Techniques de l'Ingénieur*, no. BM 5008, Paris, France, 2005.

[PEL 97] PELLETIER J.-L., *La maîtrise des risques dans les PME: 10 exemples d'applications pratiques de la sûreté de fonctionnement*, CETIM-ISdF, Senlis, France, 1997.

[PEY 03] PEYRAS L., Diagnostic et analyse de risques liés au vieillissement des barrages – Développement de méthodes d'aide à l'expertise, Civil Engineering Thesis, Blaise Pascal University – Clermont II, LERMES et CEMAGREF, Clermont-Ferrand, France, 2003.

[PIL 03] PILLAY A., WANG J., "Modified failure mode and effects analysis using approximate reasoning", *Reliability Engineering and System Safety*, vol. 79, 2003.

[PRE 09] PREVINFO, "Outils en management QHSE" [online], 2009, available at: http://www.previnfo.net (accessed 27/04/2009).

[SCT 03] SCTRICK L., GOUSSY., "Document unique concernant les risques biologiques – Risque d'accident exposant au sang (et autres liquides biologiques)" [online], September 2003, available at: http://anmteph.chez-alice.fr/docuniqaes.pdf (accessed 27/04/09).

[TAL 04] TALON A., BOISSIER D., CHEVALIER J.-L., HANS J., "A methodological and graphical decision tool for evaluating building component failure", *CIB World Building Congress (CIB WBC 2004)*, Toronto, Canada, 2004.

[TAL 06] TALON A., Evaluation des scénarios de dégradation des produits de construction, Civil Engineering Thesis, Blaise Pascal University – Clermont II, CSTB & LGC, Clermont-Ferrand, France, 2006.

[VIL 88] VILLEMEUR A., *Sûreté de fonctionnement des systèmes industriels*, Eyrolles, Paris, France, 1988.

[ZWI 96] ZWINGELSTEIN G., *La maintenance basée sur la fiabilité*, Hermès, Paris, France, 1996.

[ZWI 99] ZWINGELSTEIN G., "Sûreté de fonctionnement des systèmes industriels complexes", *Techniques de l'ingénieur*, no. S 8250, Paris, France, 1999.

Chapter 3

Application to a Hydraulic Civil Engineering Project

3.1. Context and approach for an operational reliability study

This case study illustrates an operational reliability study through its application to a hydraulic civil engineering project. Methodological developments are discussed in terms of an industrial study that the authors carried out on a flood protection system [PEY 06a]. Note that today operational reliability studies, in the form of hazard studies, are required in France under the laws relating to the safety of large hydraulic installations (dams and dykes) (edict on safety of hydraulic installations, dated 11 December 2007).

Following catastrophic events on 3 October 1988, a number of hydraulic installations were constructed around the town of Nîmes (France) with the aim of providing flash flood protection in times of flooding. The installation considered here relates to a catchment area to the east of Nîmes, covering a surface area of 16 km² (Figure 3.1). It consists of two upstream flood prevention dams with a capacity of 90,000 m³, two downstream flood prevention dams with a capacity of 400,000 m³, several kilometers of open channels and buried collectors that permit flow between the dams, and a range of other civil works (openings, grates, bypass channels, sand catchers, etc.).

Chapter written by Daniel BOISSIER, Laurent PEYRAS and Aurélie TALON.

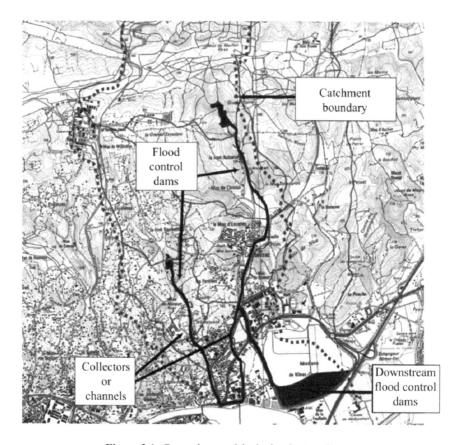

Figure 3.1. *General map of the hydraulic installation*

An evaluation of the operational reliability of such a system involves the three phases described in the preceding chapters: functional analysis and failure mode analysis (section 2.1), construction of scenarios (sections 2.1 to 2.4) and scenario-based criticality analysis (section 2.5).

For the purposes of the study, a panel of experts was formed consisting of three senior engineers specializing in dam engineering, a hydraulic engineer and a hydrologist, and it was chaired by an operational reliability specialist. This panel was consulted at the various different stages of the study, and in particular during the qualitative analysis, in order to determine the technological failure modes of the installation, and during the quantitative analysis for expert evaluation of subjective probabilities for technological failures, in cases where feedback was lacking.

3.2. Functional analysis and failure mode analysis

3.2.1. *Functional analysis of the system*

In the system under study, the hydraulic installation consisted of a succession of linear structures (buried collectors and open-air flood channels), punctuated by generic civil works (dams, flow division structures, gratings, etc.).

Three levels of granularity were used to describe the installation: the system as a whole (the hydraulic installation), the individual structures within the system (dams, collectors) and the components of each structure (sluices, spillways, filters etc.).

The geographical contours of the hydraulic installation were defined, and this then naturally led on to consideration of the external environment that it interacts with (Table 3.1).

Type	External media
Water environment	– upstream part of the rural catchment areas surrounding the hydraulic installation – downstream part of the urban catchment areas surrounding the hydraulic installation – geology, geological engineering
Environment close to the installation	– private residences, apartments and urban structures – rural zone, fields, cultivation, woodland, wasteland – vegetation, chokepoints, sediment and silt – airfields, railways – highways: minor roads, national roads and motorways

Table 3.1. *External environment interacting with the installation (extract)*

External functional analysis can be used to obtain its main functions (representing the primary purpose of the system) and its limiting functions (representing the responses of the system to environmental stresses):

– buffering water flow and flooding from upstream rural and urban catchment areas;

– resisting external environmental stresses: it must not become obstructed by any sort of obstruction/sediment/deposition, nor collapse or degrade, etc.

Internal functional analysis can then be used to study the internal behavior of the system. This consists of a structural analysis: the system is broken down into generic civil engineering structures or works, geographically distributed around the installation (Figure 3.2).

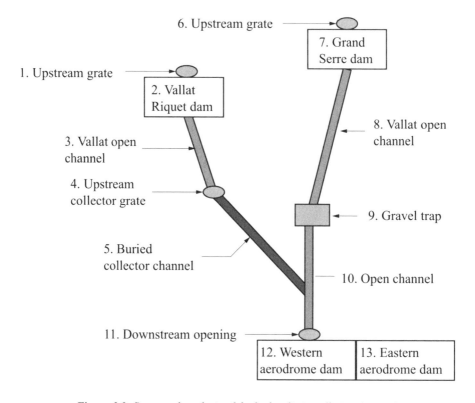

Figure 3.2. *Structural analysis of the hydraulic installation (extract)*

Such a study determines the interactions between the structures and the external environment, and finishes with the functions performed by each structure.

At the end of the functional analysis, an understanding will have been developed of the operation of the system, the functions of each structure and their interactions with the environment.

This is the starting point for the failure mode analysis.

3.2.2. *Failure mode analysis, and effects*

For each structure we must determine its failure modes, their causes and their possible effects. This investigation takes place in a systematic procedure by application of the FMEA method (section 1.3), adapted to civil engineering systems [PEY 06b], [TAL 06]. FMEA was applied to the various structures within the hydraulic installation, and its conclusions are listed in Table 3.2.

Structure	Failure mode	Causes	Effects
Flood protection dam	Does not buffer flood wave, or does so inadequately	– sluice/grating obstructed – outflow obstructed – outflow exit obstructed – downstream water level too high (failure of dissipater)	– premature filling of the reservoir with limited buffering
	Technological failure, does not resist internal erosion: – in the embankment – in the foundation – along the outflow	– erosion due to piping effects in the embankment foundation or along the outflow channel	– dam failure due to internal foundation erosion or erosion along the outflow channel

Table 3.2. *FMEA applied to a flood protection dam (extract)*

The failure mode analysis provides an appreciation of all the potential failure modes of the structures that make up the hydraulic installation. This systematic analysis can be used to examine all the failure modes of the system and the elements that it is made up of. The FMEA analysis is then subjected to expert examination, with the aim of only retaining the most relevant failure modes, in other words those that can realistically be expected to occur (including ones associated with particularly rare events). This part of the study involves individual interviews with experts as well as collective discussions leading to an expert review of all the structures.

Based on the outcome of the FMEA analysis, the experts can draw conclusions on the failure modes of the works that should be considered further, their causes and their possible effects.

In this application, the effects are considered in terms of the downstream flow from the hydraulic installation, without considering the consequences of those flows on the environment that may be flooded by them.

Finally, a list is obtained of the operational modes that can reasonably be anticipated for the works within the system, of both their nominal performance and their degraded performance.

Combining the operating modes then gives all the possible configurations of the hydraulic installation (Table 3.3).

Structure	Possible behavior and configuration
Sluice and outflow of the Vallat Riquet dam	- nominal operation - degraded operation: sluice 50% obstructed
Sluice gate on the Vallat Riquet dam	- nominal operation - degraded operation: gate stuck open - degraded operation: gate stuck closed
Conduit channel to aerodrome dams	- nominal operation - degraded operation: channel 50% obstructed

Table 3.3. *Operating configurations for various structures (extract could represent many of the dams shown in Figure 3.2)*

3.3. Construction of failure scenarios

The information required for constructing failure scenarios was obtained during the previous system analysis stage: the realistic failure modes and the corresponding configurations of the works. The failure scenarios for the system can be modeled using the event tree method.

Under this method, the sequence of events in the tree is built up in an inductive manner, starting from the initiating event and continuing until the final events.

The use of the event tree method is easy once a complete system analysis and FMEA analysis have been completed: it involves combining successive failure modes of different structures within the hydraulic installation, while respecting the laws of cause and effect in the chronology of the failures (Figure 3.3).

The initiating event in this application is systematically linked to a hydraulic event that produces a flood wave whose intensity depends on the return period T (where T is the inverse of the annual frequency) of the hydrological event.

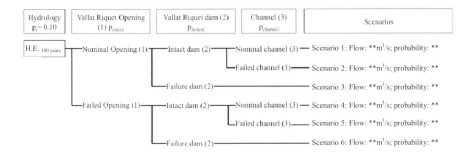

Figure 3.3. *Representation of failure scenarios
using the event tree method*

Successive individual events are the ones obtained through FMEA and its analysis by the panel of experts; they correspond to operational configurations of the hydraulic installation.

Scenario construction may prove challenging in the case of complex systems consisting of many different components, and hence of multiple failure modes (Figure 3.4).

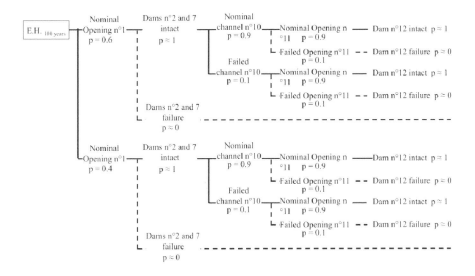

Figure 3.4. *Representation of complex failure scenarios*

3.4. Scenario criticality analysis

Scenario criticality analysis for the hydraulic installation involves a probabilistic approach for evaluating the probabilities of occurrence of individual hydrological events, combined with an expert approach for estimating the probabilities of occurrence of technological failures within the structures.

The analysis of the consequences of the various scenarios is quantified in terms of the water flow rate into the environment outside the network, and is obtained by hydraulic modeling of the installation. It therefore involves the two-criteria criticality formulation (see section 2.5.1.1).

3.4.1. *Hydrological study*

A hydrological study defines the reference rainfall events corresponding to the various return periods examined in the operational reliability study. It describes the reference rainfall levels (characterized in terms of IDF: intensity/duration/frequency) and the associated hydrographs (QDF: discharge/duration/frequency) corresponding to the discharge response of the various catchment areas for each return period from a hydrological event.

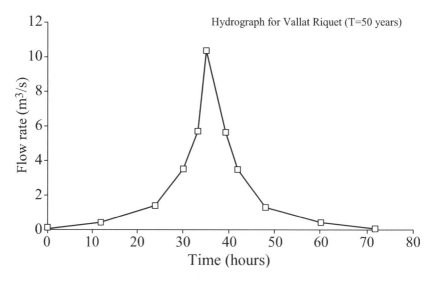

Figure 3.5. *Reference hydrograph for a specific basin
and a given return period (T = 50 years) [ARN 02]*

The computations were performed using a probabilistic hydrological method based on stochastic event simulations – the SHYREG method [ARN 02]. This method is built around a stochastic generator for hourly rainfall, calibrated for the Languedoc–Roussillon and Provence–Alpes–Côte d'Azur regions of France using data from 556 rainfall stations. The hourly rainfall histories that are generated from this model are then used to calculate the peak discharges (at 1, 2… 72 hours and for return periods from 2 to 1,000 years) on a 1 km² grid. The discharges are aggregated over each catchment basin using a statistical transfer function.

The reference rainfalls and the hydrographs are characterized by the IDF and QDF data output from the SHYREG method over the study region (Figure 3.5).

3.4.2. *Hydraulic model and quantitative consequence analysis*

Consequence analysis for a hydraulic installation considers the discharge from the system into its environment outside the water network, which represents a malfunction of the system.

In this case study, it is obtained by hydraulic modeling of the installation, which requires two different models:

– a numerical model of open surface flow (channels) and subterranean flow (collectors), using topographical data for the catchment area, and which can be used to describe the hydraulic behavior at particular points (gravel trap, diversion structures, etc.);

– a second model reproducing the hydraulic behavior of the hydraulic installations, transport within the network and flood wave buffering in dams. It is formed of computational nodes, corresponding to the input points for flood hydrographs, beyond which there may be flood protection dams, and linking channels enabling exchanges between nodes, with all these nodes being able to represent malfunctions of particular installations.

The hydraulic models simulate, for a nominal or degraded configuration of the hydraulic installation, and for a given hydrological event, the response of the system in terms of volume transported within the network and/or discharged into the external environment. They can be used to evaluate the consequences of a particular scenario.

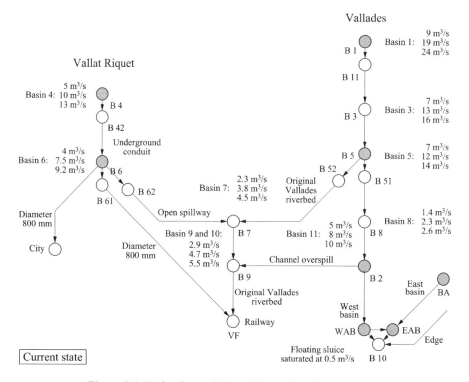

Figure 3.6. *Hydraulic modeling of flow within the installation*

3.4.3. *Evaluation of probability of technological failure*

The probabilities of occurrence of technological failures of the installations (structural failures) are evaluated through expertise, individual and collective interviews with experts, and consideration of expert verdicts. For each of the failures under consideration, each expert is questioned on their understanding of the probability of occurrence. The responses from the experts are subjected to quantitative treatment, using our proposed analysis grid, shown in Table 3.4.

The probabilities of occurrence of technological failures considered for the installations are generally determined by the hydrological event under analysis. They are evaluated as a function of the return period of the hydrological event.

Expert verdict on probability of occurrence	Quantitative treatment of expert verdict in terms of probability of occurrence
Very probable	0.60
Probable	0.40
Fairly probable	0.20
Unlikely	0.10
Very unlikely	0.01

Table 3.4. *Analysis grid for expert verdicts (from [PEY 06a])*

Finally, the scenarios are evaluated using the event tree method (Figure 3.7): the aggregations of conditional probabilities obtained at each node of the tree (corresponding to the individual failure probabilities for the installations) can be used to obtain the global probability for the scenario (the occurrence of the terminal event) and the associated consequences, expressed in terms of flow discharged into the environment, possibly in terms of areas flooded or even number of victims.

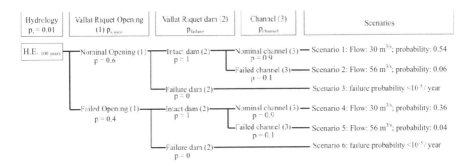

Figure 3.7. *Quantitative evaluation of failure scenarios using the event tree method (data used purely as an example)*

The dams within the hydraulic installation are the subject of a separate evaluation. This is partly due to their much greater reliability compared to the other components in the installation, and partly due to their complex failure methods that require detailed analysis on the scale of the individual components of the dam. Because of this, it is necessary to break down the dams into their elementary components and analyze their failure modes individually (Figure 3.8).

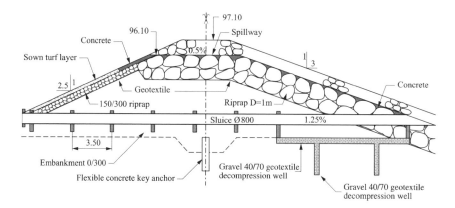

Figure 3.8. *Structural analysis of a dam within a hydrological installation*

Starting from the analysis of failure modes of the dams and their detailed review by the panel of experts, the annual probability of dam failure is evaluated in terms of the different failure modes. As an example, we will summarize the quantitative analysis of a dam in the form of a retaining embankment.

For this installation it was the overflow mechanism, evaluated in the context of various extreme hydrological events corresponding to various different return periods (T = 1,000 years to 10,000 years), that proved to be the most critical (see Table 3.5).

Failure mechanism for retaining dam (x)	Annual probability of failure
Landslide	less than 10^{-6}
Internal erosion of embankment	less than 10^{-6}
Internal erosion of foundations	less than 10^{-6}
"Piping" erosion along the outflow	less than 10^{-6}
External erosion	less than 10^{-6}
Overflow	less than 10^{-5}

Table 3.5. *Example of results of quantitative reliability analysis for a retaining embankment dam*

3.4.4. *Representing the criticality of a scenario*

Using an event tree, it is possible to class each scenario in two ways, firstly from the most probable to the least probable, and secondly from the most serious to the least serious. For each scenario, it is also possible to see the implications of failure of each installation in the probability of occurrence of the scenario.

These results can be summarized using severity–frequency curves. The criticality inherent in the hydraulic installation is shown as a function of the initiating hydraulic events associated with different return periods T (Figure 3.9).

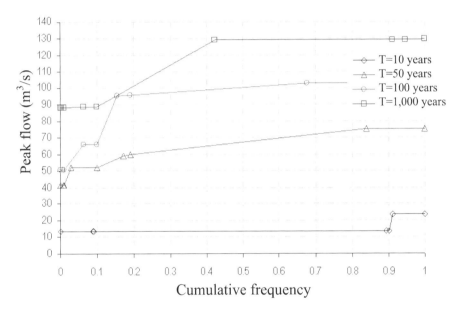

Figure 3.9. *"Severity–frequency" curves as a function of the hydrological event for different return periods*

These curves show the annual cumulative frequencies for the peak flow at a strategic point downstream of the hydraulic installation, for rainfall events with return periods of between 10 years and 1,000 years. They can be used to understand the global effectiveness of the installation by integrating the failure probabilities of each of the components of the system.

The analysis of these curves is particularly instructive. It can be shown logically that, as the consequences of failure of the installation (expressed here in terms of peak flow released into the downstream environment) become more serious, the associated annual cumulative frequency decreases. But, above all, it shows that technological failures of the installations have a higher impact on the global risk during rare hydrological events. Thus, in order to reduce the risks, it is important to protect the structures that make up the installation by taking extremely unusual rainfall events into consideration.

Furthermore, the probability of occurrence of floods and the probabilities of technological failure of the installation can be combined in order to obtain a global expression for the risk, taking into account all the different hazards and failure modes (Figure 3.10). This summarization of the risk inherent to the installation does not contain detailed information on the various failures, but it has the advantage of making it easy to appreciate the global operational reliability for the hydraulic installation, combined with the potential consequences to the environment. It typically corresponds to the results obtained in summary risk analysis performed in the industrial sector [VIL 88].

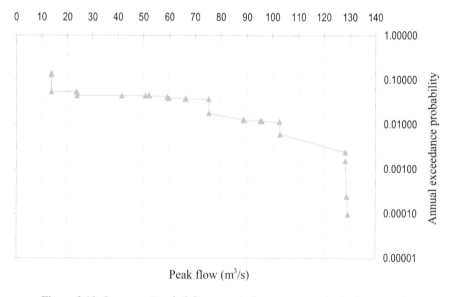

Figure 3.10. *Summary "probability–severity" curves at a point in the network*

3.5. Application summary

In the face of a complex civil engineering system, such as a hydraulic installation comprising several interconnected structures, the purpose of a reliability study is to demonstrate the most probable malfunctions and the most critical scenarios. The underlying aim is, of course, ultimately to improve the reliability of the installation and its component parts.

In this application, the most probable technological failures have been shown to be associated with the connecting components (openings, flow division structures, etc.), which might at first glance appear to be subsidiary to the main installations (the flood control dams); however, our studies have shown that failures of these structures are nevertheless liable to jeopardize the entire operation of a hydraulic installation. Our study has enabled us to propose improvements and to evaluate their benefits in terms of gains in operational reliability.

Through exhaustive examination of the environment of the system during the functional analysis, the study enabled us to identify a hitherto rather neglected risk associated with flows coming from a nearby catchment area which may, under certain conditions, interfere with the functioning of particular structures. In response to this, corrective measures have been proposed and their effectiveness evaluated.

Finally, analysis of the most probable scenarios leading to a significant risk or a major degradation in the effectiveness of the hydraulic installation has made it possible to propose preventative measures that can improve the reliability of the system. These measures of course include improvements to the civil engineering projects that comprise the installation, with the aim of significantly improving their reliability in the face of exceptional hydrological events that they prove particularly vulnerable to. Other measures consist of preventative inspection and maintenance actions that can be used to ensure the continued effectiveness of the dams in times of flood.

3.6. Bibliography

[ARN 02] ARNAUD P., LAVABRE J., "Coupled Rainfall Model and Discharge Model for Flood Frequency Estimation", *Water Resources Research*, vol. 38, no. 6, 1075, 2002.

[PEY 06a] PEYRAS L., ROYET P., SALMI A., SALEMBIER M., BOISSIER D., "Etude de la sûreté de fonctionnement d'un aménagement hydraulique de génie civil: application à des ouvrages de protection contre les inondations de la ville de Nîmes", *Revue Européenne de Génie Civil*, vol. 10, p. 615–631, Hermès, Paris, France, 2006.

[PEY 06b] PEYRAS L., ROYET P., BOISSIER D., "Dam ageing diagnosis and risk analysis: Development of methods to support expert judgement", *Canadian Geotechnical Journal*, vol. 43, p. 169–186, 2006.

[TAL 06] TALON A., CHEVALIER J.-L., HANS J., Failure Modes Effects and Criticality Analysis research for and application to the building domain. Publication CIB no. 310. Working commission 80: Service life methodologies – Prediction of service life for buildings and components, International council for building research studies and documentation, Rotterdam, The Netherlands, 2006.

[VIL 88] VILLEMEUR A., *Sûreté de fonctionnement des systèmes industriels*, Eyrolles, Paris, France, 1988.

Heterogeneity and Variability of Materials: Consequences for Safety and Reliability

Introduction to Part 2

In the majority of cases, failures in construction projects result from the combined effects of several deleterious factors. These factors may be related to how particular actions (unanticipated or underestimated) are taken into account. They may also be related to the way material capabilities and limitations have been taken into account: insufficient surveying, poor understanding of novel materials, or the use of materials under conditions that are outside of their design specification. Calculation or modeling errors can also be a source of problems. These factors are often compounded by a lack of care during construction and inadequate oversight. Every one of these factors has implications on the level of safety inherent in the final structure.

Consequently, in order to reduce the level of risks (and to quantify it as precisely as possible) it is essential to specify the role played by each factor. Ultimately, aside from cases where static equilibrium is lost, a failure will occur because the effects of external sources acting on the materials exceed what the materials are capable of tolerating, whether due to exceptional forces or because the properties of the materials are insufficient.

Part 2 considers how the variability and heterogeneity of soil or construction materials can be taken into account, and their consequences for structural safety.

Chapter 4 considers the characterization of uncertainty in geotechnical data, the identification of sources of uncertainty, data classification (aberrant, censored and sparse), and the statistical representation and modeling of these data. Chapter 5 gives some estimates on material variability (mean or characteristic values) and includes a case study of a geostatistical study in an urban environment. Chapter 6 illustrates these concepts with the example of a shallow foundation footing, the reliability of which is investigated: first the available models of bearing capacity are described, then the effects of soil variability on variations in bearing capacity and the safety of the footing is investigated; finally, the structure of the spatial correlation is taken into account in order to study its influence on the safety of the footing.

Chapter 4

Uncertainties in Geotechnical Data

4.1. Various sources of uncertainty in geotechnical engineering

Civil engineering projects all share the one similarity that they use a wide variety of construction materials (concrete, steel, wood, masonry, polymers, etc.) and that they are built on the ground, whether this be natural terrain or artificial foundations. A lack of knowledge or a poor understanding of the properties of these materials is a source of risk. In particular, the properties of the ground material, which has an impact on the strength of a structure, depends on the entire history of its formation (tectonics, erosion, transport, sedimentation, etc.). The ground also includes "heterogeneities" in the form of continuous variations (rigidity varying with depth, cohesion, Poisson coefficient, friction, etc.), or discrete variations (fractures, watercourses, "hard" points or areas of weakness). A thorough appreciation of geotechnical considerations is thus crucial in order to minimize the uncertain character of predictions of the behavior of the ground or an installation that is built upon it.

In the first section in this chapter, we will discuss different categories of terrain, in terms of "degree of uncertainty" [FAV 04], which increases as a result of sources of heterogeneity and variability. We will then define various types of data and sources of uncertainty.

Chapter written by Denys BREYSSE, Julien BAROTH, Gilles CELEUX, Aurélie TALON and Daniel BOISSIER.

4.1.1. *Erratic terrain, light disorder and anthropogenic terrain*

Jean-Louis Favre distinguishes various degrees of uncertainty within the ground [FAV 04]:

– highly erratic terrain, with strong heterogeneities, whose geometry and properties are highly inhomogeneous (with fractures, areas of weakness or "hard points" etc.). In this case, the behavior of installations is more likely to involve "accidents", in other words extreme (and improbable) values rather than middle-of-the-road values. The most appropriate way of addressing this situation involves correct surveying of the soil in order to detect these accidents, by combining geological information with geophysical or geotechnical data. Deterministic representation of "accidents" and their consequences is the logical approach to take;

– lightly disordered terrain, with regular stratification, for which each layer has properties that only vary to a limited extent, but which are significantly different from the neighboring layers. In this case, we can establish a mechanical model of the terrain, combining geological and geotechnical information, and exploit it in a statistical manner. Probabilistic approaches prove particularly suitable in this context. One of the areas where researchers are most active is the possibility of soil liquefaction. The susceptibility of the soil to liquefaction (in other words, the loss of all bearing capacity in the event of a dynamic action such as an earthquake) can be estimated using measurements obtained through penetrometry. The method then involves processing geotechnical data and modeling their spatial variability in order to build up a map of the risk of liquefaction [HIC 05];

– anthropogenic terrains, whose construction process (embankments, dams, etc.) is well understood, and for which there may be a wide selection of test data. Statistical methods can also, of course, be applied to this situation as well.

4.1.2. *Sources of uncertainty, errors, variability*

Geotechnical surveying does inevitably introduce a range of errors. In addition to the limited scope of the survey area and the number of tests (sampling density), testing errors may result in unsuitable or poorly interpreted tests [JCS 06]. These errors include:

– observation errors: measurement errors associated with the test device (calibration, zeroing, etc.), temporal errors (due to variations between the time of measurement and the time when the data is exploited, during the service lifetime of the structure);

– surveying errors, which arise from the fact that the volumes investigated are not necessarily appropriate (surveying to insufficient depth, for example);

– uncertainties within materials can also be classified in terms of aleatory and epistemic uncertainties.

Aleatory (or statistical) uncertainties can be ascribed to natural (intrinsic) variability of materials. Ostensibly comparable samples will in fact have different responses. These differences may have natural explanations in the case of soil (the properties of which result from geological erosion, transport and sedimentation processes) or wood (whose properties result from the growth process). They may alternatively be explained by the construction process: during the fabrication of concrete for example (and depending on the sophistication of the plant equipment), tolerances are specified on the composition of each batch but the actual water content of the aggregates cannot be precisely known. A standard exists [CEN 06] to impose a maximum value on the water to cement ratio, with a tolerance of 0.02 for each load of concrete produced (verification requires systematic availability of weight certificates for each load of concrete used in the installation, whether the concrete is mixed on site or in a factory).

In addition to aleatory uncertainties there are also *epistemic (or systematic) uncertainties*, associated with a lack of understanding, which may for example result from having taken limited samples. The physical or financial means available to acquire information are of course limited and, as a result, a small number of sample measurements may not necessarily be representative of the population. The statistics of standard surveys indicate, for example, that the ratio of the volume of samples taken to the volume of the foundation loaded by a dam is equal to around 10^{-5} [FAV 99]. The properties of the whole foundation must therefore be estimated from limited sampling, when it may be extremely heterogeneous. This is a major source of uncertainty in geotechnical engineering, where the variability of material is high.

An additional uncertainty factor results from the measurement itself: for a value of X which is assumed to be the "true" value of the parameter under study, the measurement protocol (probe, electronic device, etc.) provides a measured random value x for which X is only, in the case of unbiased measurements, the expected value. In the absence of bias, an infinite number of repeated measurements should cause x to tend towards the value of X. In reality, the estimation of X is less accurate if the degree of repeatability of the measurements is lower. For example, it can be estimated that the coefficient of variation in the measurement of a plate bearing test, a standard geotechnical measurement, is of the order of 30%. For such a coefficient of variation, a measurement of x < 0.5 X or x > 1.5 X could be obtained in 10% of cases! In practice, the degree of repeatability of the measurement is in general unknown, and this source of uncertainty cannot be distinguished from natural variability, despite its fundamentally different origin.

In addition to individual uncertainties, multiple biasing factors may also exist: a poor physical understanding, or imperfections in theoretical models for the interpretation and use of measurements, inappropriate simplifications of the actual scenario, tests affected by systematic errors, etc. These aspects are not discussed here, and we will restrict ourselves to discussing the effects of natural variability.

There are therefore a range of arguments justifying the modeling of material properties as random variables. Table 4.1 gives the estimated orders of magnitude for the variability of various common properties of materials.

Material	Property	Coefficient of variation (*cv*)	References
Concrete	Compression strength	4 to 10% (laboratory) 17% 8 to 11% (HPC on site) 5 to 7% (HPC in laboratory)	[IND 93] [TOR 50]
	Compression strength	9 to 25% (fiber-reinforced concrete, depending on mean resistance)	[BRE 96]
Wood	Longitudinal modulus of elasticity	23% (high grade spruce)	[ROU 93]
	Longitudinal tensile strength	33% (high grade spruce)	[ROU 93]
	Bending strength	24 to 41% (timber, depending on quality of wood)	[REN 97]
Soil	Density	5 to 10%	[JCS 06]
	Undrained cohesion	23% (fine compacted soil) 10 to 40% (clay)	[KOU 98] [PHO 96]
	Coefficient of permeability	68 to 90% (saturated clay)	[DUN 00]

Table 4.1. *Order of magnitude of variability in mechanical properties of construction materials (HPC: High Performance Concrete)*

Model errors can also be included under the umbrella of epistemic errors in the case where a model is used to interpret the results of a test that does not directly give the value of the property of interest [DNV 07]. This is the case, for example, in the Ménard pressuremeter modulus, which stems from an analysis of the results of pressuremeter tests and their interpretation (see Chapter 5). The mean values that are generally taken for the Poisson coefficient are generally $v \sim 0.3$, except in the case of saturated soils where $v \sim 0.5$. Other parameters that may have uncertainties associated with them [MAG 07] include the void index e, the water content w ($cv \sim 20\%$), the dry volumic weight γ_d, water volumic weight γ_w, grain volumic weight γ_S ($cv \sim 5\%$), the shear resistance parameters ($cv \sim 30\%$), etc. It is thus useful to take into account the relationships that may exist between some of these parameters.

4.1.3. Correlations between material properties

The correlations between properties can be expressed in terms of a range of equations [MAG 07]:

– exact mathematical equations, including those representing the physical state of soils:

$$e = n/(1-n); \quad \gamma = \gamma_d (1 + w)$$

$$w = e\gamma_w S_r / \gamma_S$$

$$\gamma = (\gamma_S \gamma_W + \gamma_S \gamma_d - \gamma_d \gamma_W)/\gamma_S$$

where S_r is the degree of saturation and n is the porosity;

– evolution equations as a function of depth: the stresses increase with depth. In the case of homogeneous deposits of fine soils whose state has stabilized, the effective stresses, consolidation pressures, moduli and resistances also increase with depth;

– empirical equations (or "correlations"): in order to simultaneously analyze the values of several properties of the same soil, it is generally assumed that the desired equations are linear. This assumption does not exclude the existence of nonlinear relationships between soil properties: the random variables linked by linear equations may be nonlinear functions of the soil properties (logarithms, power functions, exponentials, etc.), which imbues this type of linear analysis with a great flexibility. Statistical studies can, for example, give various empirical correlations [MAG 07] between what the Eurocodes refer to as measured parameters and derived parameters (for example the void ratio in a clay, and its permeability), or between different measurements: maximum measured pressure and cone resistance to a static penetrometer, or, alternatively, static and dynamic cone resistances, etc.

4.2. Erroneous, censored and sparse data

In order to evaluate the failure scenarios for civil engineering installations, it is useful to collect and process the available experimental data. The characterization of these data is, however, particularly challenging in the case of erroneous, censored or sparse data.

4.2.1. *Erroneous data*

A series of measurements of a property sometimes reveal the presence of erroneous values or *outliers*, whose probability of occurrence would appear to be extremely low if the distribution of properties were to follow a regular distribution. This is, for example, the case when a batch exhibits properties that are much lower than what would be expected, or when a sample core indicates that the characteristics of the terrain are very different to those nearby.

"Measurement errors" are often invoked as an explanation for the presence of such values, but they may also be the result of accidents, either technological ones (a bad batch, for example), or geological ones. In that case, even though the outlier may be significantly different from the rest of the population of data points, it may nevertheless be crucial since it indicates a potential weakness that could be the cause of a later failure. For example, [LEM 99] studied the consequence of *low-strength events* corresponding to the occurrence of a very weak value of concrete strength for a particular batch. The difficulty lies in processing this information in an appropriate manner: should they be integrated into the body of the data set? Should they be eliminated? Should they be given special treatment? In such a way, very high coefficients of permeability at the surface of a concrete facing could be explained in terms of structural micro-cracking, rather than by a strong porosity that would reduce the durability of the material. The tendency is to reject such results, arguing that the aim is to identify the properties of the material, but is this not then taking a risk, if such cracking does exist?

Many methods exist that can be used to test such aberrant points from a statistical point of view, and these can easily be found in the literature. Consequently, we will not discuss them in this book. These methods include the Student test, the Dixon test [FAV 04], [PLA 05] or, alternatively, Principal Component Analysis (PCA) or Factorial Correspondence Analysis (FCA) [BEN 73], not to mention the "engineer's instinct". These methods must be used in conjunction with expert insight into the possible origins of these values (potential measurement problems, variability that could be explained in terms of material characteristics, etc.).

[MAG 00] showed the extent to which the apparently incongruous character of a localized region of soil, with very contrasting properties relative to the immediate environment, may disappear following detailed expert geological consideration in conjunction with appropriate data processing: he mentions a case where the analysis of geotechnical data confirmed the presence of a region (an ancient alluvial plain) whose properties were clearly distinct from its immediate environment. On another scale, analysis of the variation in air permeability of concrete in a structure results in a set of outlying values if the experimental device is placed in a region where a macro fissure is present. Should these values be retained? If they are retained, we should probably take into account the bimodal structure of the statistical distribution (regions with and without fissures). If they are eliminated, this move must be justified, for example by the limited extent of these fissures which are not expected to compromise the global properties of the structure.

4.2.2. Bounded data

Data may sometimes only provide partial information. Thus, in terms of reliability, it is often the case that data for a property x of the material (but also for a lifetime, see section 4.3) are bounded and hence contain little information. Recall that the property x is bounded:

– on the right side (or left, respectively) by c if it is known for certain that $x > c$ (or $x < c$);

– by an interval if it is known for certain that $a < x < b$.

Two classic examples of bounded data are:

– lifetimes measured by fatigue: when the number of repetitions of the action leading to failure is too high, the test may be halted, in other words $N_{failure} > N_{halt}$ (upper bound on the period of correct operation); and

– data obtained through observations carried out during scheduled operations (SO) at particular intervals. It is then known that $t(SO_{i-1}) < t_{failure} < t(SO_i)$: bounding by interval.

4.2.3. Sparse data

There are many dangers of small sample sizes and data carrying little information for a statistical analysis: strong variability in estimates, high sensitivity to atypical values, exaggeration of contrasts, etc. Thus, when faced with such sample sets, it can be useful to bolster the reliability of the analysis in a number of different ways:

– we could abandon the analysis. This is a logical attitude but often one that is over-cautious;

– it may be possible to bypass the difficulty. This is a common attitude in the context of statistical learning or data mining. For example, in the case of missing or inadequate data, interpolation or extrapolation could be used. Then, however, there is a need to attempt to estimate the quality of the values obtained. It is also possible to generate additional fictitious values that conform to the statistical distribution of the available data, which is one way of generating a large number of new samples and to perform statistical analysis. This technique, known as *bootstrap resampling*, has for example been used to perform reliability analyses in geotechnical engineering, in a situation where there are only a small number of survey data points;

– regularization of statistical estimates. This attitude is often the best choice. It involves the introduction of external information (real or artificial information, or information based on expert opinion) to supplement the data. Thus, this strategy often naturally leads to the adoption of a Bayesian approach.

Finally, we should mention a class of techniques that are intended to enhance the contribution of certain points or suppress the influence of aberrant points within the sample set. This type of approach falls into the classification of robust statistical analysis [HAM 86] and it is, by definition, particularly useful in the presence of aberrant or poor quality data. In all cases, a preliminary sensitivity analysis can be used to identify the most useful data points (see Figure I.2 in the Introduction to this book).

In the rest of this second part of the book, we will assume that the data are "complete" or, in other words, that "there are a sufficient number of data points". For example, in order to estimate the mean and standard deviation of a parameter, Eurocode 7 [CEN 07] recommends at least five samples. Conversely, in order to have an idea of the statistical distribution of a parameter, no less than 20 samples are often required. Bayesian inference, particularly useful for this type of analysis, is introduced in Part 4 of this book. An example of an analysis of bounded and sparse data is given in section 4.3.

4.3. Statistical representation of data

4.3.1. *Notation*

Let $x = (x_1..., x_i...,x_N)$ be a vector of N values for a physical property (cohesion, modulus of elasticity, friction angle, etc.). This parameter x is known at n points, obtained through a measurement process such as, for example, event counting (earthquakes, rockfalls, etc.) over a given period, or obtained from sample measurements. In order to model this discrete parameter, it is helpful to interpret the measured data (statistical studies) and to define suitable hypotheses. For this, a histogram is often employed. The data are sorted in increasing order. They are then grouped into classes: the spread of measurements $x_{max} - x_{min}$ is divided into r intervals I_j of equal size. For each interval, the absolute frequency (or number of measurements) n_j and relative frequency $f_j = n_j/n$, which defines the histogram, are determined, such that $\sum_{j=1}^{n} f_j = 1$. The cumulative frequency $F_i = \sum_{j=1}^{i} f_j$ then gives the cumulative frequency curve.

After having chosen r classes[1], the standard estimators are determined:

– central estimators, such as the mean: $\bar{x} = \sum_{i=1}^{n} f_i \cdot x(i) = \frac{1}{n} \sum_{i=1}^{n} x_i$, the mode $m_0 = x(j)$, most observed value, in other words $f_j = \max(f_i)_{i=1,...n}$, the median $Me = x(j)$, the value such that 50% of the observed values are below this value (in other words $\sum_{i=1}^{j-1} n_i < n/2$ and $\sum_{i=1}^{j} n_i \geq n/2$);

– spread estimators: variance within a sample of n known values:

$$s^2 = \sum_{i=1}^{n} f_i \cdot (x(i) - \bar{x})^2 = \frac{1}{n} \sum_{i=1}^{n} (x_i - \bar{x})^2$$

and $s'^2 = \frac{1}{n-1} \sum_{i=1}^{n} (x_i - \bar{x})^2$ the unbiased estimator of the complete population;

– shape estimators (coefficients of asymmetry or flatness);

– variation estimators (coefficient of variation $cv = s/\bar{x}$).

These parameters are evaluated by statistical estimation (samples, measurements) and/or according to prescribed rules (for example, Eurocode [CEN 07]). Section 4.4.5 illustrates these concepts using resistance measurements for samples of concrete. Chapter 5 will extend this initial analysis in order to estimate the variability in material characteristics.

---

1 The Sturges formula [STU 26] may be useful: $r \sim 1 + 3.22 \log_{10} n$. For example, if $n \sim 10$, $r \sim 4$. It is also often considered that there should be at least five elements in most classes, and no gaps.

4.3.2. *Spatial variability of material properties*

In certain soils, properties measured more than 50 cm apart may have no correlation between them. In other soils, on the other hand, the properties are largely the same over several meters, or even several tens of meters. The *autocorrelation distance* of a property reflects the degree of similarity of this property between two points as a function of the distance that separates them. Several authors have stated values for the autocorrelation distances for soil cohesion, based on test measurements in the laboratory and *in situ*. We note that the scale of horizontal fluctuation (5–50 m) is much greater than the vertical scale (0.5–5 m). The quantity of information on autocorrelation distances is limited; to establish it precisely requires a large number of measurements, and no "general rule" is currently known that is applicable to all types of soil. The values for autocorrelation distances used in geotechnical calculations must therefore be validated through a detailed investigation to identify these distances. The value of the estimator for the fluctuation scale is strongly influenced by the sampling interval (the interval over which the survey has been performed) [FEN 99]. Thus, during a measurement campaign, in order to obtain a good value for the correlation values, it is necessary to perform measurements separated by distances smaller than the autocorrelation distance δ. The maximum sampling distance that is often recommended for a soil in the vertical direction, is of the order of $\Delta v = \delta v/2$, whilst in the horizontal direction it is more like $\delta h/8 < \Delta h < \delta h/4$ [POP 95].

A tool widely used in geo-statistics to characterize spatial variability is the variogram. The next chapter describes this and illustrates it with an analysis of piezometric measurements in an urban environment. In other structured construction materials (concrete, wood), the concept of the autocorrelation distance is of course relevant, but there is very little information available on plausible values for this distance.

4.4. Data modeling

Data collected in order to evaluate failure scenarios for civil engineering works may take a wide range of different forms: experimental data, expert opinion, data from mechanical models, sounding tests, etc. Similarly, a wide range of tools are available to model these data. In the next section we will discuss probabilistic and possibilistic approaches; focusing on possibilistic description in the next section.

4.4.1. *Probabilistic and possibilistic approaches*

Probabilistic data modeling makes use of probability theory. A distinction is made between discrete random variables and continuous random variables. In this book we will not discuss the discrete case (binomial distributions, Poisson distributions, etc.). X is defined as a continuous random variable if the set of possible values it can take consists of a continuous interval of values. Such a variable can be defined in terms of the probability that X will take a value in all intervals $[x, x + h]$, where h is real.

This is expressed in terms of the partition function of X, written F_X and defined by $F_X(x) = p [X < x]$. This can be used to calculate $p [x \leq X < x + h] = F_X(x + h) - F_X(x)$, where $p [X = x] = 0$. If the partition function for X is continuous, we have $F_X(x) = \int_{-\infty}^{x} f_X(u).du$, where the probability density function f_X is characterized by a number of parameters (mean μ, standard deviation σ, etc.). The coefficient of linear correlation $\rho_{XY} = Cov[X,Y]/(\sigma_X \sigma_Y)$ is used to describe correlations in the variation of two random variables X and Y, where σ_X, σ_Y represent the standard deviations of X and Y and $Cov[X,Y]$ their covariance.

Probability theory is very widely used to describe the propagation of uncertainties by way of a mechanical model or to estimate a risk. However, not all types of data are applicable to such an approach. For example, expert opinion [TAL 06], [CUR 08] may prove difficult to quantify. Expert opinion can be defined as the combination of experimental data and experience: for example, if there is a brownish stain on a rendered facade (fact) then this may be due to the presence of humidity and microorganisms (experience) [TAL 06].

Possibilistic approaches have more limited application and are awkward to apply, but they can be used to represent the main types of experimental data [HRY 06], [MAS 06] or expert opinion within the same formalism, as fuzzy subsets.

Associated with any expert opinion is an uncertainty which is completely unrelated to the uncertainty resulting from the (intrinsic) variability of the estimated parameters. Fuzzy subsets, which have their roots in the theory of fuzzy logic, can be used to formalize expert opinion by taking into account this uncertainty and inaccuracy. A trapezoidal fuzzy subset A of X is described by several characteristics (Figure 4.1):

– its kernel: $Kern(A) = \{x \in X \mid \mu_A(x) = 1\}$;

– its support: $Supp(A) = \{x \in X \mid \mu_A(x) > 0\}$;

– its height: $h(A) = \sup\limits_{x \in X} \mu_A(x)$ (a fuzzy subset is normalized if $h(A)=1$);

– its cardinality: $|A| = \sum\limits_{x \in X} \mu_A(x)$;

– its α-cut: $A_\alpha = \{x \in X | \mu_A(x) \geq \alpha\}$.

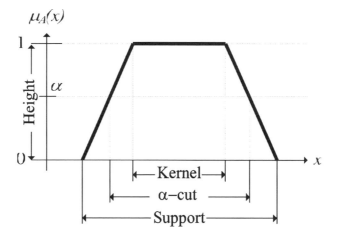

Figure 4.1. *Characteristics of a trapezoidal fuzzy subset A of X*

For further information, readers are referred to Chapter 9, which discusses an application in greater detail (in a study into the evacuation time for a building following the outbreak of a fire).

4.4.2. *Useful random variables (Gaussian, Weibull)*

The Gaussian, or normal, random variable, written X, is one of the most widely used random variables, and it is represented by the following probability density:

$$f_X(x) = \frac{1}{\sigma\sqrt{2\pi}} \exp\left(-\frac{1}{2}\left(\frac{x-\mu}{\sigma}\right)^2\right) \qquad [4.1]$$

This distribution is thus entirely represented by two parameters: its mean μ and its standard deviation σ. It is also written as $N(\mu, \sigma^2)$. If a normal distribution is to be used, it is useful to confirm that the variables under study are continuous and are fairly numerous (>20–30 data samples).

In addition, the random variable U, the reduced centered normal distribution, is defined by the following probability density:

$$f_U(x) = \frac{1}{\sqrt{2\pi}} \exp\left(-\frac{x^2}{2}\right)$$

[4.2]

This distribution, written as $N(0,1)$, is indispensible in statistics and all types of computational code. It can be found in the form of the reduced centered normal distribution table [VER 07]. This distribution is used in many places in this book; in particular, it is used to represent the mean distribution of material properties. It is, however, important to be aware of the potential problems that may be encountered if a normal distribution is used to represent a random variable that must remain positive for physical reasons (density, modulus of elasticity, dimension of a component, etc.), particularly when the variance is high (greater than 0.3, for example). In that case, there may be a significant probability of generating negative values, and it is useful to bound the distribution with a minimum value, an action which therefore alters the observed distribution.

The Weibull distribution[2] is one of three distributions (along with the Gumbel and Fréchet distributions) that can be used to describe the statistical distribution of "extreme" phenomena such as the resistance to breaking of a material. Note that $x=(x_1..., x_n)$ for n measured resistances. This distribution has the form:

$$F_W(x) = 1 - \exp(-[(x-a)/\eta]^\beta)$$

where β, a and η are three real numbers ($\beta > 0$ and $\eta > 0$), which are known as the shape, location and scale parameters. The probability of exceeding a given value can also be expressed by writing $P[x] = 1 - F_W(x)$. A two parameter Weibull distribution refers to the case where a is zero, and a three parameter Weibull distribution refers to the more general situation. Determination of the parameters from experimental data is simple in the case of a two parameter distribution (if $a = 0$) by rewriting the expression for the distribution such that:

$$\ln(-\ln[1-F(x)]) = \beta \ln x - \beta \ln \eta$$

---

2 Waloddi Weibull, a Swedish engineer, worked between 1937 and 1939 on the statistical distribution of the elastic limits and mechanical resistances of steel. He showed that "his" distribution was equally suitable for describing the resistance of steel, the height of adult British males or the size of bean seeds [WEI 51].

β and η can then be identified by plotting the values for the partition function on a diagram where $ln\ x$ is used for the horizontal axis and $ln\ (\ -\ ln\ [\ 1 - F(x)\ ]\ )$ for the vertical axis. If a is non-zero, it is less simple to identify the parameters, and a minimization procedure (least squares) must be used. The probability density then becomes:

$$f_W(x) = \frac{\beta}{\eta}\left(\frac{x-a}{\eta}\right)^{\beta-1} e^{-(\frac{x-a}{\eta})^\beta}\ ;\ \ x > 0\ ;\ \beta > 0, \eta > 0 \in \mathbf{R}^2 \qquad [4.3]$$

Weibull's statistical model has various advantages [LAM 07]:

– it has a simple form. In the case of a two parameter distribution ($a = 0$), the two other parameters can be estimated very easily from the partition function. This is why it is easy to interpret, and is used by a wide range of researchers and engineers;

– it can be used to describe the statistical distribution of resistances to breakage of a large number of samples, under simple force conditions (single-axis traction, for example);

– it can be used to estimate a wide range of distributions (extreme wind speeds, flood wave heights, etc.) and can be used to estimate the reliability of a component.

4.4.3. *Maximum likelihood method*

In order to extract useful information from a sample of data $(x_1..., x_n)$ belonging to $\mathbf{R}^p$, statistical modeling involves assuming that the data are drawn from an unknown probability function of density $f(x)$. It is usually assumed that this density function belongs to a parametric family that is suitable for representing and summarizing the observed phenomenon. Thus, we assume that $f(x) = f(x;\ \theta)$, where θ is an unknown vector parameter that must be estimated from $(x_1..., x_n)$.

Suppose, for example, that the data $(x_1..., x_n)$ represent the resistance to breakage of n samples. If these resistances follow a normal distribution of mean μ and variance σ^2, the parameter to be estimated is $\theta = (\mu, \sigma^2)$. If the resistances follow a Weibull distribution, then the parameter to be estimated is $\theta = (\beta, \eta)$. A widely-used approach for estimating the parameters of a statistical model is *maximum likelihood estimation* [VER 07]. The likelihood of the parameter associated with the data (complete in this case) is written:

$$L(\theta) = \prod_{i=1}^n f(x_i; \theta)$$

This function contains all the information provided by $(x_1..., x_n)$ about the parameter θ. The maximum likelihood method involves estimating θ using:

$$\hat{\theta} = \arg\max_\theta L(\theta)$$

This estimator has useful properties when n is large compared to the dimensions of θ: it converges to the optimum value of the parameter and has minimal variance. Conversely, in the case where n is small, the likelihood estimator may prove unstable (see [BAC 98] for illustrations of this issue for situations relevant to industrial experience feedback). We have, for example, already given the maximum likelihood estimator $\hat{\theta} = (\bar{x}, s^2)$ for the case of a normal distribution, where $\bar{x}$ and s^2 are defined, in section 4.3.1. In the case of a two parameter Weibull distribution $\hat{\theta} = (\hat{\beta}, \hat{\eta})$, its maximum likelihood is obtained by iteratively solving the following equations:

$$\frac{1}{\hat{\beta}} + \frac{\sum_{i=1}^{n} \ln(x_i)}{n} - \frac{\sum_{i=1}^{n} x_i^{\hat{\beta}} \ln(x_i)}{\sum_{i=1}^{n} x_i^{\hat{\beta}}} = 0 \text{ and } \hat{\eta} = \left[\frac{\sum_{i=1}^{n} x_i^{\hat{\beta}}}{n} \right]^{\frac{1}{\hat{\beta}}} \qquad [4.4]$$

When the problem becomes rather more complex, which is commonly the case for industrial experience feedback data, it is necessary to use several different models in parallel in order to achieve reliable and useful statistical inference. The choice of model is thus an important phase in a good statistical analysis.

We will not consider various contemporary statistical techniques that are useful for validating the choice of model.

4.4.3.1. Data adequacy testing: χ^2 test

The χ^2 test, amongst others, gives a measure of the adequacy between a frequency histogram and a probability density function, characterized by a distribution and m parameters. We can, for example, consider a normal distribution or a Weibull distribution, characterized by a mean and a standard distribution ($m = 2$), or alternatively by ($m = 3$) parameters. The spread is divided into r intervals of equal size (Sturges rule [STU 26]: 5 for $n = 20$, 6 for $n = 30$). Let p_i be the theoretical probability that a value will belong to class i, n_i the absolute observed frequency and $\upsilon = r - 1 - m$ the number of degrees of freedom. The D^2 statistic is then calculated, and this is compared to the value $\chi^2_{\upsilon,\alpha\%}$ of the Pearson variable, for a fixed confidence level $\alpha\%$ (which is often 1–5%). If $D^2 > \chi^2_\upsilon$, the candidate distribution is rejected. Table 4.2 gives various values of $\chi^2_{\upsilon,\alpha\%}$ for common values of α and υ [VER 07]:

$$D^2 = \sum_{i=1}^{r} \frac{(n_i - n\,p_i)^2}{n\,p_i} \; ; \; D^2 > \chi^2_\upsilon \text{ where } P(\chi^2_\upsilon > \chi^2_{\upsilon,\alpha\%}) = \alpha\% \qquad [4.5]$$

$\alpha(\%)$	1	5	1	5	1	5	1	5
ν	3	3	4	4	5	5	6	6
$X^2_{V,\alpha\%}$	11.3	7.81	13.3	9.49	15.1	11.1	16.8	12.6

Table 4.2. *Common values of* $\chi^2_{\upsilon,\alpha\%}$ *extracted from the Pearson distribution table*

This test can be used for $n > 20$, but works best for $n > 30$ [FAV 04].

4.4.3.2. *Likelihood ratio test*

In order to choose between two models, M0 and M1, whose parameter spaces are nested, the likelihood ratio test is used:

$$\Lambda = \frac{L(\hat{\theta}_0)}{L(\hat{\theta}_1)}$$

If we assume that the data result from the distribution $f(x, \theta_0)$, then $-2 \ln \Lambda$ asymptotically (as n tends to infinity) follows a distribution of χ^2 with dim θ_1 - dim θ_0 degrees of freedom.

4.4.3.3. *Penalized likelihood criteria*

Penalized likelihood criteria are used to choose a model from a set of models that are not necessarily nested. The most widely used [PER 08] are the AIC (*Akaike Information Criterion*) and the BIC (*Bayesian Information Criterion*). For a model M, these are written:

$$AIC(M) = -2L(\hat{\theta}_M) + 2\dim(\hat{\theta}_M)$$

and:

$$BIC(M) = -2L(\hat{\theta}_M) + \dim(\hat{\theta}_M)\ln(n)$$

These criteria to be minimized are obtained using asymptotic arguments and are optimal when the size of the data sample is large enough.

Thus, statistical inference from probabilistic models is based on the convergence of observed frequencies to the probabilities that they estimate. When the user has a wide range of data that provides definite and precise information, there is generally a broad choice of available statistical models, and the user is free to use relatively complex methods to analyze the data.

In such cases, statistical analysis offers a powerful tool for extracting the most important and relevant elements in order to solve difficult problems and make robust decisions. The difficulties begin when there is only a small amount of available data, or the data does not contain much information about the phenomenon under study. The latter case will be considered in the next section.

4.4.4. *Example: resistance measurements in concrete samples*

Table 4.3 gives $n = 27$ measurements of compressive resistance of concrete samples, performed on control samples that were cast at the same time as a construction. These were arbitrarily divided into $r = 6$ classes. For each class, the medians x_i and the relative frequencies f_i, $i=1,...,$ $n = 6$ were determined, and the mean was found to be 31.2 MPa, the standard deviation $s = 1.7$ MPa (variance $s/\bar{x} = 6\%$), and the median and mode estimated to be around 31 MPa.

Classes (MPa)	27–28.5	28.5–30	30–31.5	31.5–33	33–34.5	34.5–36
Effective n_i	1	6	9	6	3	1
Relative frequency f_i and cumulative frequency (F_i)	0.04 (0.04)	0.23 (0.27)	0.345 (0.615)	0.23 (0.845)	0.115 (0.96)	0.04 (1)

Table 4.3. *Measurements of compressive resistance of concrete samples divided into classes of resistance, along with relative and cumulative frequencies*

Consider the test of a normal distribution to model this distribution. The normal distribution is described by $m = 2$ parameters: a mean $\mu = \bar{x}$ and a standard deviation $\sigma = s$. For a class i, of frequency n_i, the probability $p_i = f_X(x_i)$ is defined by equation [4.1].

Using equation [4.1], the adequacy between this frequency histogram and the normal distribution is estimated using the statistic $D^2 = 5.9$ defined in equation [4.5], by assuming that $f_X(x_i)$ such that $X \sim N(31 \text{ MPa}; 1.7 \text{ MPa})$. It can be confirmed that $D^2 < X^2_{v=3.5\%} = 7.8$ (Table 4.3). Thus the normal distribution tested here is not rejected, with parameters $\hat{\theta} = (31;1.7)$. The Weibull distribution with three parameters is not rejected either, for which we obtain $D^2 = 6.9$, where $p_i = f_W(x_i)$ defined in equation [4.3] such that $\hat{\theta} = (\hat{a}, \hat{\beta}, \hat{\eta}) = (25;4;7)$. Conversely, in this example the trial of a two parameter Weibull distribution is rejected ($D^2 = 156$), for $\hat{\theta} = (\hat{\beta}, \hat{\eta}) = (24;32)$.

4.5. Conclusion

The primary aim of this chapter was to summarize the preliminary approach taken in all studies that take into account the variability and heterogeneity of material properties: identifying the uncertainties, gathering and classifying the available data, and then modeling these data using a model that is consistent with the data. This chapter included several samples which serve as an introduction to the discussions that will follow.

4.6. Bibliography

[BAC 98] Bacha M., Celeux G., Idee E., Lannoy A., Vasseur D., *Estimation de modèles de durées de vie fortement censurés*, Eyrolles, Paris, France, 1998.

[BAU 07] Baudrit C., Couso I., Dubois D., "Joint propagation of probability and possibility in risk analysis: Towards a formal framework", *International Journal of Approximate Reasoning*, vol. 45, p. 82–105, 2007.

[BEN 73] Benzecri J.P., *L'analyse des données*, vol. 2, Dunod, Paris, France, 1973.

[BRE 96] Breysse D., Attar A., "Comportement du béton renforcé de fibres métalliques: hétérogénéité et statistique", *Coll. Int. francophone sur les bétons renforcés de fibres métalliques*, Toulouse, France, 1996.

[CEN 04] CEN, Eurocode 0: Basis of Structual Design, Brussels, 2004.

[CEN 06] CEN, EN 206-1 Part 1: Method of Specifying and Guidance for the Specifier, Brussels, 2006.

[CEN 07] CEN, Eurocode 7: Geotechnical Design - Part 1: General Rules, Brussels, 2007.

[CLE 95] Clement J.-L., Du matériau à la structure, Cours de Matériaux, ENS Cachan, France, 1995.

[CUR 08] Curt C., Evaluation de la performance des barrages en service basée sur une formalisation et une agrégation des connaissances, PhD thesis, University of Clermont-Ferrand II, France, 2008.

[DEG 96] DEGROOT D.J., "Analyzing spatial variability of in situ properties. Uncertainty in the Geologic Environment, from Theory to Practice", *Special ASCE Publication*, no. 58, p. 210–238, 1996.

[DNV 07] DET NORSKE VERITAS, Statistical Representation of Soil Data, Recommended Practice, DNV-RP-C207, 2007.

[DUB 06] DUBOIS D., "Possibility theory and statistical reasoning", *Computational Statistics & Data Analysis*, vol. 51, p. 47–69, 2006.

[DUN 00] DUNCAN J.M., "Factors of safety and reliability in geotechnical engineering", *J. Geot. Geoenv. Eng.*, ASCE, no. 126, vol. 4, p. 307–316, 2000.

[FAV 99] FAVRE J.-L., "L'aléa terrain dans les barrages en terre. Entretiens avec P. Londe et J.J. Fry", *Préparation du colloque IREX Risque et Génie Civil*, 1999.

[FAV 04] FAVRE J.-L., *Sécurité des ouvrages – risques: modélisation de l'incertain, fiabilité, analyse des risques*, Ellipses, Paris, France, 2004.

[FEN 99] FENTON G.A., "Estimation for stochastic soil models", *Journal of Geotechnical and Geoenvironmental Engineering*, ASCE, no. 125(6), p. 470–485, 1999.

[GID 01] GIDEL G., Comportement et valorisation des graves non traitées calcaires utilisées en assises de chaussées souples, PhD thesis, University of Bordeaux 1, France, 2001.

[HAM 86] HAMPEL F.R., RONCHETTI E.M., ROUSSEEUW P.J., *Robust Statistics: The Approach Based on Influence Functions*, Wiley, New York, USA, 1986.

[HIC 05] HICKS M.A., ONISIPHOROU C., "Stochastic evaluation of static liquefaction in predominantly dilative sand fill", *Géotechnique*, no. 55, vol. 2, p. 123–133, 2005.

[HRY 06] HRYNIEWICZ O., "Possibilistic decisions and fuzzy statistical tests", *Fuzzy Sets and Systems*, vol. 157, p. 2665–2673, 2006.

[IND 93] INDELICATO F., "A statistical method for the assessment of concrete strength through microcores", *Mat. Str.*, RILEM, no. 26, p. 261–267, 1993.

[JCS 06] JCSS, Probabilistic model code – section 3.6, Soil Properties, version revised by J. BAKER and E. CALLE, 2006.

[KOU 98] KOUASSI P., Comportement des sols fins compactés: application aux remblais et aux ouvrages en terre, PhD thesis, University of Bordeaux 1, France, 1998.

[LAM 07] LAMON J., *Mécanique de la rupture fragile et de l'endommagement: approches statistiques et probabilistes*, Hermes, Lavoisier, Paris, France, 2007.

[LEM 99] LEMING M.L., "Probabilities of low-strength events in concrete", *ACI Str. J.*, p. 369–377, 1999.

[MAG 00] MAGNAN J.P., "Quelques spécificités du problème des incertitudes en géotechnique", *Rev. Fr. Géotechnique*, no. 93, p. 3–9, 2000.

[MAG 07] MAGNAN J.P., "Corrélations entre les propriétés des sols", *Techniques de l'ingénieur*, www.techniques-ingénieur.fr/dossier/correlationséntre_les_propriétés_des_sols.

[MAS 06] Masson M.H., Denoeux T. "Inferring a possibility distribution from empirical data", *Fuzzy Sets and Systems*, vol. 157, p. 319–340, 2006.

[MAT 08] Matthies H.G., "Structural damage and risk assessment and uncertainty quantification", *NATO-ARW Damage Assessment and Reconstruction After Natural Disasters and Previous Military Activities*, Sarajevo, Bosnia and Herzegovina, 5–9/10/2008.

[PHO 96] Phoon K.K., Kulhawy F.H., "On quantifying inherent soil variability, Uncertainty in the geologic environment", *ASCE Specialty Conference, Madison, WI, Reston*, p. 326–340, 1996.

[PLA 05] Planchon V., "Traitement des valeurs aberrantes: concepts actuels et tendances générales", *Biotechnol. Agron. Soc. Envir.*, 9, p. 19–34, 2005.

[POP 95] Popescu R., Stochastic variability of soil properties: data analysis, digital simulation, effects on system behaviour, PhD thesis, University of Princeton, USA, 1995.

[REN 97] Renaudin P., Approche probabiliste du comportement mécanique du bois de structure, prise en compte de la variabilité biologique, PhD thesis, University of Bordeaux 1, France, 1997.

[ROU 93] Rouger F., De Lafond C., El Quadrani A., "Structural properties of french grown timber according to various grading methods", *CIB W18 Conference*, Athens, Greece, 1993.

[SAG 00] Sageau J.F., "Le suivi préventif des structures", *Risques et Génie Civil*, Presses ENPC, p. 541–552, 2000.

[STU 26] Sturges H., "The choice of a class interval", *J. Amer. Statist. Assoc.*, vol. 21, p. 65–66, 1926.

[TAL 06] Talon A., Evaluation des scénarios de dégradation des produits de construction, Civil Engineering Thesis, Centre scientifique et technique du bâtiment – division environnement, Clermont-Ferrand, France, *Produits et Ouvrages Durables et Laboratoire Génie Civil*, 2006.

[TAN 07] Tannert C., Elvers H.D., Jandrig B., "The ethics of uncertainty. In the light of possible dangers, research becomes a moral duty", *EMBO Reports*, no. 8, vol. 10, p. 892–896, 2007.

[TOR 50] Torroja E., *Sur le coefficient de sécurité dans les constructions de béton armé, reissued in Fifty Years of Evolution of Science and Technology of Building Materials and Structures*, F.H. Wittmann, RILEM, 1997.

[VER 07] Verdel T., Décision et prévision statistique, School of Mines, Nancy, France, 2007, www.thierry-verdel.com.

[WEI 51] Weibull W., "A Statistical Distribution Function of Wide Applicability", *Journal of Applied Mechanics*, p. 293–297, 1951.

Chapter 5

Some Estimates on the Variability of Material Properties

5.1. Introduction

In this chapter, we will consider, through a number of examples, the issue of estimating the mean and the representative values of a material property in the presence of variability. We will discover the role played by sample size (number of measurements) and the level of confidence that is expected. We will then consider the question of spatial correlation of properties, which can be modeled in order to reproduce, through simulation, the behavior of materials and structured heterogeneous media. A range of civil engineering materials have structures of this type, often for physical, geological or biological reasons. This is obvious in the cases of soils and rocks, wood or concrete, for example. This spatial structuring may also result from the fabrication process: batch production, layer-based construction, compacting, etc.

5.2. Mean value estimation

5.2.1. *Sampling and estimation*

Suppose that measurements of simple compressive strength are performed on three batches of concrete, giving the following results: 53.8 MPa, 62.7 MPa, 49.3 MPa. What can we conclude about the properties of the batch of concrete that these samples are taken from?

Chapter written by Denys BREYSSE and Antoine MARACHE.

If we assume that the strength r_c can be described by a random variable, we have three instances of this random variable, R_{c1}, R_{c2} and R_{c3}, which we can use to determine an empirical mean $\bar{r}_c = (R_{c1} + R_{c2} + R_{c3})/3 = 55.3$ MPa. Obviously three different samples, selected at random from the same population, would have yielded different results. The question we must therefore ask is how representative the result we have obtained is: to what extent is $\bar{r}_c$ a good estimate of the true mean, which we will write as $\mu(r_c)$?

The empirical mean gives an estimate of the true mean value for the population. This estimate converges towards the exact solution as the sample size becomes very large. In fact, there is a specific probability that the exact value lies within an interval $I = [\bar{r}_c - \Delta, \bar{r}_c + \Delta]$ about the empirical mean: by writing the probability in the form $1-\alpha$ (α is the risk that we are wrong) we can define a confidence interval for the estimate [RUE 89].

The equality of the estimate is obtained by calculating the confidence interval (at the $1-\alpha$ level). It can be shown that the extent of the confidence interval depends on:

– the standard deviation $\sigma(r_c)$ of the distribution of the variable;

– N, the number of samples: as the number of samples increases, the risk of error decreases;

– the desired confidence level (if a high confidence level is required), in other words if there is a need to have a very small chance of error in the estimate, then a larger sample size is required.

The expression for the interval exploits the characteristics of the convergence of the estimated value towards its exact value (central limit theorem). It is as follows:

$$I = [\bar{r}_c - f(\alpha)\sigma(r_c)/\sqrt{N}, \bar{r}_c + f(\alpha)\sigma(r_c)/\sqrt{N}] \qquad [5.1]$$

The exact expression varies depending on whether the standard deviation $\sigma(r_c)$ is known or not:

– if it is known, the function $f(\alpha)$ is given by the values of the normal distribution:

$$f(\alpha) = z_{\alpha/2} = \phi^{-1}(\alpha/2)$$

Table 5.1 gives some values of $z_{\alpha/2}$ for common values of $1-\alpha$;

– if the standard deviation is not known (which is more commonly the case), an uncertainty on the estimate of the variance must be added to that on the estimate of the mean.

$1 - \alpha$	0.999	0.995	0.99	0.95	0.90	0.80
$z_{\alpha/2}$	3.29	2.807	2.576	1.96	1.645	1.282

Table 5.1. *Common values of $z_{\alpha/2}$ taken from the normal distribution table*

The expression for the interval is then written as:

$$I = [\bar{r}_c \ -t_{\alpha/2} \ s(r_c \ )/\sqrt{N}, \ \bar{r}_c \ +t_{\alpha/2} \ s(r_c \ )/\sqrt{N} \ ] \qquad [5.2]$$

where $s(r_c)$ is the estimated standard deviation (or empirical standard deviation), which replaces the true mean, and where $t_{\alpha/2}$ is a coefficient obtained from the Student's t-distribution. Table 5.2 gives values of $t_{\alpha/2}$ as a function of the number of degrees of freedom (the number of samples minus one). Note that as the number of samples increases, the value of $t_{\alpha/2}$ approaches that of $z_{\alpha/2}$.

Thus, if these equations are applied to the data at the start of this section, the empirical standard deviation is equal to 6.82 MPa[1]. If we assume that the variance is unknown, then for two degrees of freedom and at the following confidence levels we have:

– level $1 - \alpha = 50\%$, $t_{\alpha/2} = 0.8165$, which gives I = [52.1; 58.5],

– level $1 - \alpha = 90\%$, $t_{\alpha/2} = 2.920$, which gives I = [43.8; 66.7].

Both intervals are centered on the empirical mean, and the numerical values confirm that, if we want to increase the confidence level, then the width of the interval also increases. Conversely, if the confidence level is reduced then this is represented with a "risk of error" in the estimate.

In our example, the values of the resistance of the three samples were in reality "drawn at random" from a population with a Gaussian distribution, with a mean of 50 MPa and a standard deviation of 5 MPa. It can therefore be seen that the interval of the estimate at a confidence level of 50% does not contain the true value of the mean, which is consequently overestimated. Figure 5.1 shows the result of a

---

1 The standard deviation of the population is evaluated based on the sample (Eurocode 0, Appendix D) [CEN 03].

numerical simulation illustrating the process of convergence to the mean, as the number of measurements in the sample set is increased.

Num DoF	Confidence level 1 - α				
	0.99	0.95	0.90	0.75	0.50
1	63.6559	12.7062	6.3137	2.4142	1.0000
2	9.9250	4.3027	2.9200	1.6036	0.8165
3	5.8408	3.1824	2.3534	1.4226	0.7649
4	4.6041	2.7765	2.1318	1.3444	0.7407
5	4.0321	2.5706	2.0150	1.3009	0.7267
6	3.7074	2.4469	1.9432	1.2733	0.7176
7	3.4995	2.3646	1.8946	1.2543	0.7111
8	3.3554	2.3060	1.8595	1.2403	0.7064
9	3.2498	2.2622	1.8331	1.2297	0.7027
10	3.1693	2.2281	1.8125	1.2213	0.6998
12	3.0545	2.1788	1.7823	1.2089	0.6955
15	2.9467	2.1315	1.7531	1.1967	0.6912
20	2.8453	2.0860	1.7247	1.1848	0.6870
25	2.7874	2.0595	1.7081	1.1777	0.6844
30	2.7500	2.0423	1.6973	1.1731	0.6828
40	2.7045	2.0211	1.6839	1.1673	0.6807
50	2.6778	2.0086	1.6759	1.1639	0.6794
75	2.6430	1.9921	1.6654	1.1593	0.6778
100	2.6259	1.9840	1.6602	1.1571	0.6770
1,000	2.5807	1.9623	1.6464	1.1510	0.6747

Table 5.2. *Standard values of $t_{\alpha/2}$ taken from the table of the Student distribution for five different confidence levels, as a function of the number of degrees of freedom*

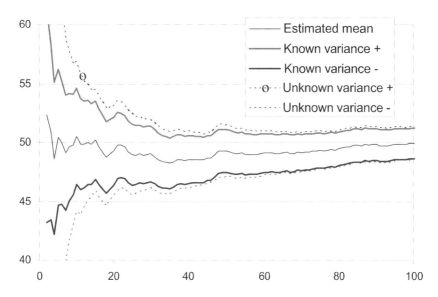

Figure 5.1. *Simulation of convergence of the estimate for the mean, as a function of the sample size – in other words the number of measurements (99% confidence level)*

5.2.2. *Number of data points required for an estimate*

The theory used for the estimation can also be used to quantify the number of samples required to obtain an estimate of the properties with a previously-determined level of confidence. For this, we simply need to refer back to equation [5.1], which gives the confidence interval (for a confidence level $1-\alpha$) as a function of N.

We will apply this approach to the case of estimating the speed of propagation of ultrasonic waves in concrete (which is a common method of indirectly and non-destructively measuring the modulus of elasticity of the material). If we know that the velocity V can be treated as a random variable with a mean equal to 4,200 m/s and standard deviation 100 m/s, how many measurements should we perform in order to estimate the mean to +/- 50 m/s at a confidence level of 95%? What if we increase the confidence level to 99%? The precision of the estimate is $\Delta = +/- z_{\alpha/2} \, \sigma(V)/\sqrt{N}$. We therefore have a minimum value of $N = [\, z_{\alpha/2} \, \sigma(V)/\Delta\,]^2$.

With a confidence level of 95%, we have $z_{\alpha/2} = 1.96$, $\Delta = 50$ and $\sigma(V) = 100$, so that $N = (1.96 \times 100/50)^2 = 16$ measurements. If the required confidence level is 99%, we have $z_{\alpha/2} = 2.58$ and $N = 27$ measurements.

5.3. Estimation of characteristic values

5.3.1. *Characteristic value and fractile of a distribution*

Eurocode 0 [CEN 03] states that the characteristic value of a material or product is the value that has "a prescribed probability of not being attained in a hypothetical unlimited test series. This value generally corresponds to a specific fractile of the assumed statistical distribution of the particular property of the material or product".

We therefore often want to obtain representative values of these properties, which would be "safe" values (reasonably pessimistic), in other words ones corresponding to a lower fractile of the distribution (in this case of quantities such as resistance). The term "characteristic value" is then used for these material properties.

Fractile	10%	5%	0.1%	10%	5%	0.1%
No. samples	Known standard deviation			Unknown standard deviation		
1	1.81	2.31	4.36	-	-	-
2	1.57	2.01	3.77	3.77	7.73	-
3	1.48	1.89	3.56	2.18	3.37	-
4	1.43	1.83	3.44	1.83	2.63	11.40
5	1.40	1.80	3.37	1.68	2.33	7.85
6	1.38	1.77	3.33	1.56	2.18	6.36
8	1.36	1.74	3.27	1.51	2.00	5.07
10	1.34	1.72	3.23	1.45	1.92	4.51
20	1.31	1.68	3.16	1.36	1.76	3.64
30	1.30	1.67	3.13	1.33	1.73	3.44
∞	1.28	1.64	3.08	1.28	1.64	3.08

Table 5.3. *Values of coefficient k_n for three fractiles*

In practice, this series of unlimited tests is not just hypothetical: it is impossible. Estimating the characteristic value (corresponding to a fixed lower fractile) requires estimating both the empirical mean value $m(x)$ and the dispersion of the property (empirical standard deviation $s(x)$ or variance $var(x)$). The quality of the estimate then depends, for a chosen level of risk, on the size of the sample set. Eurocode 0 (Appendix D (Informative): Dimensioning Assisted by Experiment) states how these fractiles may be estimated. The generic expression for a characteristic value X_k has the form:

$$X_k = m(x) - k_n s(x) = m(x) (1 - var(x)) \qquad [5.3]$$

where the value of the coefficient k_n depends on the level of confidence (Appendix D of Eurocode 0 [CEN 03] states that the estimate corresponds to a confidence level of 75%), on the number of tests, the desired fractile and on whether the standard deviation is assumed to be known (or that a realistic upper bound for it is known) or not. Table 5.3 summarizes the values given for k_n.

5.3.2. Example: resistance measurements for wood samples

Flexion tests performed on six samples of wood yielded the following values (MPa):

152.2; 139.9; 152.4; 130.6; 184.1; 125.5.

Let us estimate the characteristic bending strength corresponding to the 5% fractile.

We calculate the empirical mean and the empirical standard deviation:

– with six tests: $k_n = 2.18$, $m(x) = 147.5$ MPa, $s(x) = 21.0$ MPa

It follows that:

$X_k = 147.5 - 2.18 \times 21.0 = 101.6$ MPa

Four additional tests give new results:

137.3; 158.3; 94.4; 75.3.

How do these results modify the estimate obtained from the first six tests?

We can calculate the new empirical mean and empirical standard deviation:

With ten tests:

$k_n = 1.92$, $m(x) = 135.0$ MPa, $s(x) = 31.5$ MPa.

It follows that:

$X_k = 135.0 - 1.92 \times 31.5 = 74.6$ MPa

Note that for a modest number of samples (as is common in civil engineering contexts) the estimate of the fractile is very sensitive to the individual results obtained from the samples taken. The fact that the ninth and tenth samples encountered had much smaller characteristics has significantly lowered the value of the estimated fractile.

It is not, of course, possible to determine the "true" solution experimentally. In this case, however, we can state that the ten random values were "drawn at random" from a Gaussian distribution with a mean of 120 MPa and a standard deviation of 20 MPa, which corresponds to a theoretical 5% fractile equal to 87.1 MPa.

5.3.3. *Optimization of number of useful tests*

In the previous section we considered the question of the number of tests required to achieve a particular precision. Another question to consider is the "optimal" number of tests required in order to characterize the properties of a construction material or a soil. The final cost of the construction is assumed to result from the sum of the cost of the tests (laboratory investigations, program of terrain surveying, etc.) and the cost of building the structure. We have also just seen that a more precise and more optimistic estimate of the characteristic properties is possible if more test results are available. Table 5.3 gives the value of k_n that can be taken in order to estimate the characteristic value, depending on the number of tests available. This value is higher if the number of tests is larger, since that enables us to estimate the value for a given fractile with a greater degree of confidence. A higher value for the characteristic value therefore makes it possible, during the design of the structure, to reduce the construction cost, since less material is required in the dimensioning of the structure (reduction of cross sections, masses, etc.).

We will show that the optimum number of tests depends on the level of variability within the material, by assuming that the global cost can be written in the very simplified form $C_{tot} = C_{tests} + C_{construction}$, and assuming that the standard deviation is known. We also assume that the unit cost of a test is equal to $C_{unit\ test} = 1$ U and that the cost of construction has the form $C_{construction} = 200/X_k\ U$, where U is the unit of cost. We will also assume here that the tests lead in every case to the same construction solution being selected. The cost of failure is therefore not a variable.

The expression for the global cost can be written:

$$C_{tot} = N\ C_{unit\ test} + 200/X_k$$

where the first term is an increasing (linear) function of N and the second is a decreasing function of N. The value of X_k corresponding to the 5% fractile is determined from Table 5.3, under the assumption that the standard deviation is known. The sum of these two terms will thus always possess an optimum value, representing the number of tests that results in a global minimum cost, as can be seen in Figure 5.2. We find that, as the variability increases, it makes economic sense to increase the number of tests in order to obtain a more accurate estimate of the characteristic value and reduce the design cost. The additional cost of the tests is compensated by the gains resulting from reductions in other costs.

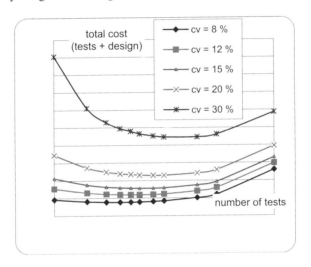

Figure 5.2. *Variation in global cost as a function of the number of tests for various degrees of material variability (example). Cv for coefficient of value*

5.3.4. *Estimate of in situ concrete mechanical strength*

The need to evaluate the mechanical resistance of concrete *in situ* in a structure presents a major challenge. A recent standard [CEN 07] suggests two ways of carrying out this evaluation.

The "reference" (direct) method involves removing samples (cores) and performing statistical analysis on the results. Depending on the number of cores N, the equations used to estimate the characteristic resistance $f_{ck,is}$ vary slightly (the subscript "*is*" stands for "*in situ*"):

– if $N > 14$: $f_{ck,is} = \min (f_{m(n),is} - k_2\ s; f_{is,\ min} + 4)$; where $s = \mathrm{Max}\ (s, 2\ \mathrm{MPa})$ and $k_2 = 1.48$;

– if $N < 14$: $f_{ck,is} = \min (f_{m(n),is} - k(N); f_{is, min} + 4)$; where $k(N) = 7$ if N lies between 3 and 6, $k(N) = 6$ if N lies between 7 and 9, and $k(N) = 5$ if N lies between 10 and 14.

In these equations, s is the empirical standard deviation and $f_{is, min}$ is the smallest of the measured values for the cores and all the strengths are in MPa.

The indirect method involves use of non-destructive testing techniques (measurement of propagation speed of ultrasonic waves, rebound measurements or pullout tests), along with a model M that is be used to determine the strength from the values of the non-destructive measurements. The standard proposes "standardized" models which must then be subjected to a calibration procedure that is used to adapt the standardized distribution to the particular concrete in the structure under study.

This calibration procedure also requires a minimum number of cores, at points where both the strength and the value of the non-destructive measurement have been identified. The calibration involves the empirical mean and standard deviation of the differences between the measured value of the strength and the value estimated using the standardized model.

The quality of the final estimate depends for the most part on three factors: the number of measurements, the quality of these measurements, and the quality of the standardized model. The calibration procedure is, however, designed to reduce the effects of an inadequate model [BRE 10].

To follow on from our discussion of statistical estimation, we will now describe the geostatistical approach for studying spatial or temporal variability.

5.4. Principles of a geostatistical study

5.4.1. *Geostatistical modeling tools*

Various tools exist to help characterize the spatial and/or temporal variability of properties. Here we will mostly consider geostatistics, focusing on those aspects that have direct consequences on reliability problems. From a geostatistical point of view, the terms "spatial" and "temporal" have the same significance, a variable sampled over time being treated using similar tools to a spatially sampled variable.

Furthermore, here we will only consider the analysis of the variability of a single variable. It should be remembered, however, that the tools described here can also be applied to the study of several correlated variables (multivariable geostatistics

[WAC 03]) and can take account of possible differences in the support of the measurements.

Geostatistics aims to reveal spatial relationships that may exist between data in order to understand the structure of such data [CHI 99], [MAT 62]. Thus, we will study a natural phenomenon distributed over space or time, known as a regionalized phenomenon. A regionalized variable, numerical in type, is used to describe the regionalized phenomenon over a bounded domain. In geostatistics, the regionalized variable is treated as a random field $Z(x)$ with realization $z(x)$. This field is assumed to be characterized by an appropriate probability space, with a mean (or expected value) of $\mu_Z(x)$, a standard deviation $\sigma_Z(x)$ and an autocorrelation function $\rho_Z(x, x')$. The field is said to be homogeneous (or stationary, in the case of a stochastic process) if it is translation invariant over space; it has the same statistical moments at all points in space. It is second order stationary if it only possesses a first order moment m_1 (mean) and a centered second order moment m_2. The autocorrelation function ρ_Z of the field thus only depends on the distance $h = x' - x$ that separates the two points with coordinates x' and x:

$$\mu_Z(x) = \mu_Z \; , \; \sigma_Z(x) = \sigma_Z \; , \; \rho_Z(x, x') = \rho_Z(h)$$

A wide range of functions can be used to describe the correlation structure [BAR 05]. Furthermore, we will consider stationary processes as ergodic: their spatial means converge towards their mathematically expected values.

The tool that enables us to study the structure of a regionalized variable, treated as a random field, is the variogram $\gamma(h)$ (or, in fact, the semi-variogram; see equation [5.4]). The variogram can be combined with other functions that can be used to characterize the variability of a variable, such as the autocorrelation function or the covariance function, but it can be shown that the variogram is a more general tool than either of these alternatives [CHI 99]:

$$\gamma(h) = \frac{1}{2} Var(Z(x + h) - Z(x)) \tag{5.4}$$

The parameter h corresponds to a distance that is regularly increased during the calculation of the variogram. We then plot $\gamma(h)$ as a function of h (Figure 5.3). A random field is stationary if its variogram is bounded. In this case, the variogram reaches a plateau whose value is equal to the variance of the variable, at a distance h known as the range (in the example shown in Figure 5.3, the plateau is equal to 1 and the range equal to 5). Beyond the range (or *correlation length*), there is negligible spatial correlation between the two points. If the variogram is unbounded, the random field is intrinsically of zero order if the growth of the variogram is less rapid than that of h^2, or of order k in the inverse case. This latter type of variogram is observed in the case of data that includes a general overall trend, which is for example the standard situation in geotechnical engineering due to weight effects (a "geostatic" trend in parameters that increase naturally with depth). In that case, we can either work with a non-stationary random field, or filter the global trend (characterized, for example, by regression) and work with the residual variable, which generally speaking offers the advantage of giving access to more detailed structural information that would have been masked by the broad trend before it was filtered.

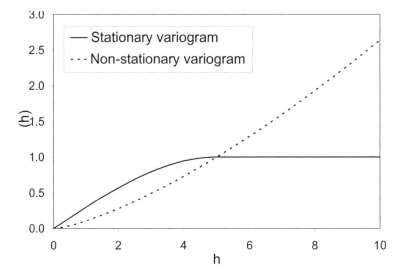

Figure 5.3. *Examples of stationary and non-stationary variograms*

Experimentally, the variogram is estimated through the following equation:

$$\hat{\gamma}(h) = \frac{1}{2N_h} \sum_{x_j - x_i = h} \left[z(x_j) - z(x_i)\right]^2 \qquad [5.5]$$

where x_i and x_j represent the positions of each point of the pair and N_h is the number of such pairs of points that are separated by h. A certain tolerance is allowed on the distance h, in order to consider the full set of possible pairs of points within the domain of the study. Generally speaking this calculation is performed up to distances h that are as large as half the largest distance between points (the size of the domain to be explored). Furthermore, it is possible either to calculate an omnidirectional variogram by taking into account all pairs of points in any possible orientation, or alternatively to calculate a directional variogram by only counting the pairs of points that correspond to a predefined direction (again, with a certain tolerance). This second mode of calculation can be used to reveal any potential anisotropy in the spatial structure of the data.

During analysis of an experimental variogram with the aim of characterizing the variability of the parameter under study, the following questions must be answered (Figure 5.4):

– is the variable stationary, and what are the values of the range and the plateau?

– are there nested structures present (structures of different scales which leave an imprint on the variogram in the form of several different ranges, in this case equal to 5 and 15)?

– is there periodicity (*hole effect*) in the data (here, a periodicity of 12.6, the value of h for which $\gamma(h)=\gamma(0)$)?

– what is the behavior of the variogram at the origin? Three classical behaviors can be identified which characterize the continuity of the field. A continuous field is characterized by a parabolic behavior of its variogram at the origin, whereas a discontinuous field shows a nugget effect (equal to 0.5 in Figure 5.4); between these two lies the possibility of linear behavior at the origin. In addition, the nugget effect may also represent a measurement error: due to such an error, two infinitely close points, whose true properties are in fact identical, exhibit apparently different properties. The nugget effect may thus provide information on strong differences in value at very short distances, or on a defect, or alternatively on a measurement error or an erroneous measurement.

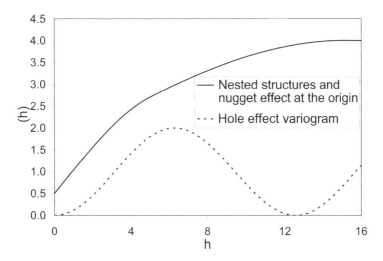

Figure 5.4. *Illustration of nested structures, the nugget effect, and periodicity, using two examples of variograms*

To apply the modeling tools described in the next section, the experimental variogram cannot be used directly. There must be an intermediate phase involving modeling of the experimental variogram. A wide range of variogram models can be found in the literature, each meeting particular constraints [CHI 99] and offering a variety of forms that can be used to fit any type of experimental variogram. The aim is to obtain an optimal fitting up to a distance h that is consistent with the extent of the neighborhood used for the modeling (the neighborhood being the set of experimental points that is used to determine the value of a variable at a position that has not been sampled).

5.4.2. *Estimation and simulation methods*

Geostatistical modeling tools can be used to estimate or simulate the value of a parameter at a location that has not been sampled, by taking into account the structural information provided by the variogram. There are numerous methods that can be used to model a continuous or discrete variable. We will not attempt to discuss all such methods here, and will focus simply on those that are of interest in the reliability of civil engineering structures. The spectrum of available modeling methods can be divided into two subsets: estimation methods and simulation methods.

Geostatistical estimation methods are all build upon a base method known as kriging[2] [MAT 62]. This is a linear, unbiased estimator that ensures minimal estimation variance. Furthermore, it is an exact interpolator (one that exactly reproduces the experimental data it is based on) and one that takes into account the position of each experimental point during the estimation process. Kriging is thus the optimal estimation method if the variogram model used is a good representation of the spatial or temporal variability of the data. Nevertheless, kriging and the methods developed from it provide a smoothed representation of reality, which is a significant drawback when considering reliability problems, problems where the extreme values are of very high importance. For this same reason, if the variable being modeled is going to be used as an input to further calculations, it can be shown that estimation methods do not deliver optimal results [CHI 99].

One way of addressing this smoothing problem is to use a simulation method that can introduce greater variability into the results than an estimation method can. Moreover, in contrast to kriging, every time a simulation is performed it will give a different result, which makes it possible to process the results in a statistical manner after a large number of simulations have been carried out (a number determined

---

2 The term *kriging* is derived from the name of a South African mining engineer, Daniel Gerhardus Krige. It also finds applications in meteorology, electromagnetism, etc.

based on the convergence criteria of the results). Geostatistical simulation methods can be divided into non-conditional simulations, whose results reproduce the spatial structure described by the variogram, and conditional simulations that, in addition to reproducing the spatial structure described by the variogram, are consistent with the experimental data and reproduce their statistical distribution. Whether conditional or non-conditional, all such simulation techniques [CHI 99] can also be described as Gaussian (turning bands, sequential Gaussian method, etc.) or non-Gaussian methods (sequential indicator simulation, for example). In the Gaussian case, the use of these methods requires that the variable being modeled follows a Gaussian distribution in order for the result to reproduce the statistical distribution of the experimental data; if the distribution is not Gaussian to begin with, an anamorphosis (or Gaussian normalization [LEM 09]) is required before the simulation can be performed.

Of the methods currently available, we feel that methods based on the use of indicators [JOU 82] are particularly promising for reliability problems. In the case of these methods, the variable $Z(x)$ under study is encoded in the form of a binary indicator variable according to a threshold (or cutoff value):

$$\begin{cases} I(x,c) = 1 & \text{if } Z(x) \le c \\ I(x,c) = 0 & \text{if } Z(x) > c \end{cases} \qquad [5.6]$$

This encoding of the variable is of particular interest for reliability problems, where the intention is generally that certain critical thresholds should not be exceeded. Thus, a set of cutoff values are defined as a function of the variable being studied. Following this encoding, the experimental variograms for the indicators are calculated. It can also be shown that the variograms for the indicators are generally less erratic than those of the raw variable, for which extreme or aberrant values can contribute to a high variability in the variogram.

A multivariate variogram model of the indicators can then be developed (with as many variables as there are cutoff values). However, in many cases this multivariate model is difficult to develop, and a simplified version of the method involves limiting oneself to a variogram model based on a single cutoff value representative of the entire system. A practical example of a sequential indicator simulation is presented in the next section.

5.4.3. *Study of pressuremeter measurements in an urban environment*

The example discussed in this section is taken from the RIVIERA project (*RIsques en VIlle, Équipements, Réseaux, Archéologie*) [THI 06]. It involves three-dimensional modeling of a geotechnical parameter in an urban environment

[MAR 09]. The parameter chosen is the pressuremeter modulus, for which it is well known that the smaller it is, the more the soil will have poor characteristics with respect to settlement. It would therefore be impossible to study the reliability of structures built upon soil susceptible to settlement without very precise modeling of the variability of the geological engineering properties of the soil.

Over the study area, the town of Pessac (in Gironde, France), there are 127 sites where pressuremeter measurements were performed, yielding 772 pressuremeter modulus values (Figure 5.5). It can be seen from the figure that the density of available data is much greater in the eastern part of the district than in the western part, where there is a much sparser density of buildings. The three-dimensional model of the pressuremeter modulus extends between the surface of the ground (an upper bound provided by a digital elevation model adjusted to the scale of the site) and the basement of the quaternary formations, which is the lower bound reconstructed by kriging under inequality constraints (a geostatistical method that ensures that all the input data is respected, whether or not boreholes have reached the lower limit: for a given borehole, if the basement of the quaternary formations is reached, the surface reconstructed by kriging corresponds to this interface, and if the drilling stops before reaching this interface, the reconstructed surface must pass below the end of this borehole).

The thickness of the terrain between these two limits varies from 0 m to 15 m. Since the details of how these limits are reconstructed are not the topic of this present book, readers are referred to [MAR 09] for further details.

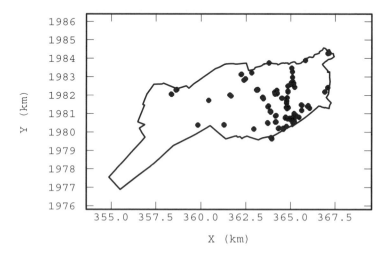

Figure 5.5. *Location of 127 pressuremeter measurements within Pessac*

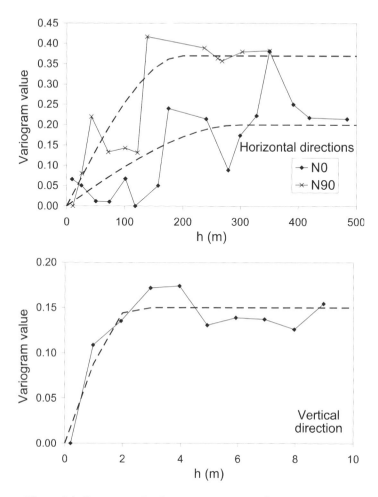

Figure 5.6. *Experimental indicator variograms and variogram models for the cutoff value of 12 MPa in the three principal directions of anisotropy (N0 = North–South, N90 = East–West)*

As we discovered in the previous section, a wide range of geostatistical methods are available for reconstructing a three-dimensional image of a geotechnical parameter. In light of our interest in reliability studies, we feel it is essential both to take correct account of extreme values of the parameter under study and to obtain probabilistic results. The modeling method we selected was the conditional sequential indicator simulation. Indicator methods require cutoff values to be defined, and these values were chosen according to geotechnical classifications of the pressuremeter modulus: 0 MPa, 4 MPa, 12 MPa and 36 MPa [PHI 00]. Since it is not easy to find a multivariable variogram model with as many variables as there are cutoff values, it is common to build the variogram model using a single cutoff value that is representative of the whole dataset. In this case, the value 12 MPa is used (in order to establish the variogram model, therefore, the pressuremeter modulus is encoded in the form of a binary variable using a threshold of 12 MPa). Experimental variograms are used to establish a model in the three directions of anisotropy (Figure 5.6) that takes account of the observed ranges and plateaux. As is often the case with geotechnical datasets, fittings in the horizontal directions are not easy, because of the high lateral variability in the parameters (even over short distances) and the rapid decrease in the number of pairs of points as the computational lag in the variogram increases.

Following the variographical analysis, one hundred conditional indicator simulations were performed, taking into account the variogram model that was obtained and an appropriate neighborhood. Figure 5.7 shows two examples of simulation results corresponding to a cross-section along the line A-B. Since each simulation result is a realization with an equal probability, by repeating the simulations it is possible to calculate the probability of encountering a given class of pressuremeter modulus within a given region. Figure 5.8 shows the regions where there is a probability of at least 80% that the pressuremeter modulus is less than 4 MPa (the size of a simulated block is in this case $25 \times 25 \times 1$ m^3).

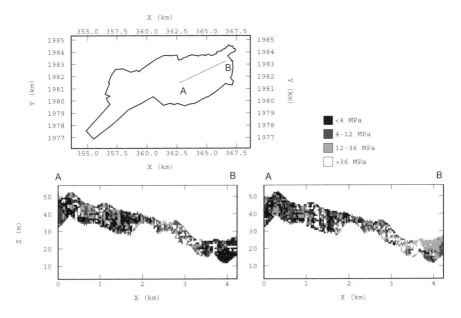

Figure 5.7. *Examples of simulation results for classes of pressuremeter modulus along the line AB*

In conclusion, we have used this example to show how the spatial variability of a property can be modeled with the help of geostatistical tools. This modeling example could be made more complex in order to improve the results, by resorting to more complex methods such as, for example, taking into account auxiliary variables (other geotechnical parameters, lithology, etc.) in the modeling of the principal variable.

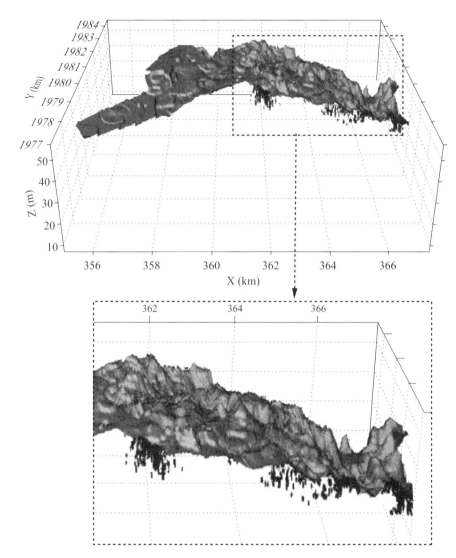

Figure 5.8. *Regions where there is a probability of at least 80% that the pressuremeter modulus is smaller than 4 MPa*

Results such as these may be used as decision-making tools, for example in the case of decisions to deploy additional boreholes, or to identify regions with particularly poor geotechnical characteristics, zones where there are direct consequences for the safety and reliability of structures that are built on the soil.

5.5. Bibliography

[BAR 05] BAROTH J., Analyse par EFS de la propagation d'incertitudes à travers un modèle mécanique non linéaire, Thesis, University of Clermont II, France, 2005, http://tel.archives-ouvertes.fr/tel-00397666/fr/.

[BRE 10] BREYSSE D., SOUTSOS M., MOCZKO A., "Non destructive assessment of concrete strength: the challenge of calibration", *Medachs Conf.*, La Rochelle, France, 2010.

[CEN 03] CEN, Eurocode 0, EN 1990 – P 06-100, Bases de calcul des structures, 2003.

[CEN 07] CEN, Assessment of in situ compressive strength in structures and precast concrete, Brussels, Belgium, 2007.

[CHI 99] CHILÈS J.-P., DELFINER P., *Geostatistics — Modeling Spatial Uncertainty*, Wiley-Interscience, Chichester, 1999.

[JOU 82] JOURNEL A.G., "The indicator approach to estimation of spatial distributions", *Proceedings of the 17th APCOM International Symposium*, p. 793–806, 1982.

[LEM 09] LEMAIRE M., *Structure Reliability*, ISTE Ltd, London and John Wiley & Sons, New York, 2009.

[MAR 09] MARACHE A., BREYSSE D., PIETTE C., THIERRY P., "Geotechnical modeling at the city scale using statistical and geostatistical tools: The Pessac case (France)", *Engineering Geology*, vol. 107, p. 67–76, 2009.

[MAT 62] MATHERON G., *Traité de géostatistique appliquée, vol. 1*, Technip, Paris, France, 1962.

[PHI 00] PHILIPPONNAT G., HUBERT B., *Fondations et ouvrages en terre*, Eyrolles, Paris, 2000.

[RUE 89] RUEGG A., *Probabilités et statistiques*, Presses Polytechniques Romandes, Lausanne, Switzerland, 1989.

[THI 06] THIERRY P., BREYSSE D., with the collaboration of VANOUDHEUSDEN E., MARACHE A., DOMINIQUE S., RODIERE B., BOURGINE B., REGALDO-SAINT BLANCARD P., PIETTE C., RIVET F., FABRE R., Le projet RIVIERA, risques en ville: Equipements, Réseaux, Archéologie, final report, BRGM/RP-55085-FR, 2006.

[WAC 03] WACKERNAGEL H., *Multivariate Geostatistics – An Introduction With Applications*, Spinger, Berlin, Germany, 2003.

Chapter 6

Reliability of a Shallow Foundation Footing

6.1. Introduction

This chapter considers the reliability of a shallow footing. We will study the effect of soil variability on the variability of the bearing capacity and safety of the foundation. Subsequently, the structure of the spatial correlation is considered in order to study its influence on the safety of the footing. Our analysis considers:

– the effects of modeling errors and variability in mechanical properties;

– a probabilistic approach that is used to characterize the statistical distribution of the bearing capacity. It highlights the need to quantify in detail the variability in mechanical properties (cohesion and friction angle);

– inclusion of correlations between these two parameters, which significantly modifies the result (based on reliable statistical data);

– the "Eurocode" semi-probabilistic approach, which appears ill-suited to taking account of correlations between different parameters; and

– spatial correlation, which is a crucial consideration for questions of differential settlement between neighboring footings, or in the context of soil-structure interaction modeling.

The example used to illustrate this approach is, as indicated, that of a foundation footing, but all the concepts considered here (model errors, dispersion of material properties, spatial correlation) can be treated in a similar manner and also apply to other problems associated with construction materials such as, for example,

Chapter written by Denys BREYSSE.

estimation of the longevity of reinforced concrete in the face of carbonation or chloride ingress. In the latter case, the properties that determine the behavior of the system are the diffusion rate, the cover thickness, the chloride ion concentration, etc. – but the analysis takes exactly the same form.

6.2. Bearing capacity models for strip foundations – modeling errors

Modeling the bearing capacity of a shallow strip foundation footing is a classic geotechnical problem. We will approach it by first analyzing the role of mechanical modeling and then of data modeling. The problem we will analyze is that of a strip foundation of infinite length and of width $B = 1$ m, buried to $D = 1$ m under the surface, in soil of weight per unit volume $\gamma = 22$ kN/m^3. We begin by comparing the values obtained using various common mechanical models, and we then consider the effects of variation in soil properties (friction angle and cohesion). We will consider the differences between correlated and uncorrelated random variable modeling, and random field modeling. We will treat two separate base configurations:

– configuration A, a purely frictional soil (friction angle $\phi' = 35°$);

– configuration B, a cohesive soil with cohesion $C' = 20$ kPa and $\phi' = 25°$.

The case of a footing on purely cohesive soil has only recently been treated in the literature [ORR 08] and we will not consider it here. We will compare the results from our calculations to the values that an engineer would use in probabilistic dimensioning using Eurocode 7 [CEN 07].

The bearing capacity of a shallow foundation can be expressed in terms of the friction angle and cohesion of the soil (mechanical characteristics that can be determined in a laboratory) using a generic formula which we can be written as:

$$q_{tot} = 0.5 \ \gamma B \ N_\gamma(\phi') + (q + \gamma D) \ N_q (\phi') + C \ N_c (\phi') \qquad [6.1]$$

where a distinction is made between "surface", "depth" and "cohesive" terms. The coefficients N_γ, N_q and N_c are all expressed as a function of the friction angle, but the expressions for these coefficients vary depending on the model that is chosen. [MAG 04] lists the most popular models, of which we consider five here.

These five models share the same coefficient N_c:

$$N_c = (N_q - 1) \cot \phi' \qquad [6.2]$$

There are two different possible expressions for N_q:

$$N_q = \exp[\,(3\pi/2 - \phi')\tan\phi'\,]/[2\cos^2(\pi/4 + \phi'/2)] \qquad \text{[TER 43]}$$

$$N_q = \exp(\,\pi\tan\phi'\,).\tan^2(\pi/4 + \phi'/2) \qquad \text{[MEY 63]}$$

It is in their definition of the coefficient N that the five models differ:

$N_\gamma = 0.5\tan\phi'.(k_p/\cos^2\phi' - 1)$, where k_p is the passive earth pressure coefficient whose value is obtained from tables   [TER 43]

$$N_\gamma = (N_q - 1).\tan(1.4\,\phi') \qquad \text{[MEY 63]}$$

$$N_\gamma = 2(N_q + 1).\tan(\phi') \qquad \text{[VES 73]}$$

$$N_\gamma = 1{,}5(N_q - 1).\tan(\phi') \qquad \text{[BRI 70]}$$

$$N_\gamma = 2(N_q - 1).\tan(\phi') \qquad \text{[CEN 07]}$$

Figure 6.1 compares the values of N_γ for the five different models, for a friction angle ϕ' varying between 15° and 40°. This confirms the nonlinear nature of the relationship, and shows the differences between models. Figure 6.2 shows the result obtained for the bearing capacity q_{tot} for a cohesion $C' = 20$ kPa, an angle ϕ' of between 15° and 40°, and dimensions $B = D = 1$ m.

The values of q_{tot} are also listed for the two reference configurations and the five models (Table 6.1). According to the "Eurocode 7" model [CEN 07], the differences (compared to the Terzaghi model [TER 43], which gives fairly large values) are no more than –7% (Meyerhof [MEY 63]) and +2.5% (Vesic [VES 73]) in configuration A, and range from –3% (Meyerhof) to +3% (Vesic) in configuration B. These differences are representative of the error in the mechanical model; the choice of model leads to an error on the estimate, which may be reduced by improving the model.

Furthermore, if there are uncertainties inherent in the model (we know which of the models is closest to reality), the discrepancies that result from repeated use of that one model forms a statistical bias (always in the same direction). Here, it is more important to note that these discrepancies are smaller than the variation in q_{tot} that results from a change of 1° in the value of the friction angle – and yet, this parameter is not known to such a high precision. We will therefore study the effects of variability in the mechanical properties of the soil in more detail.

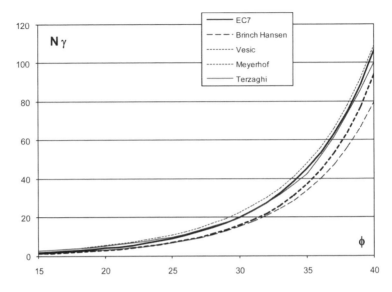

Figure 6.1. *Variation of N_γ for the five models considered,
as a function of the friction angle ϕ' (in degrees)*

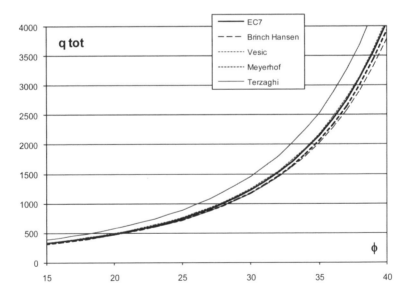

Figure 6.2. *Variation of q_{tot} (in kPa) for the five models considered (C' = 20 kPa),
as a function of the friction angle ϕ' (in degrees)*

Model	Configuration A $\phi' = 35°, C' = 0$ kPa	Configuration B $\phi' = 25°, C' = 20$ kPa
Terzaghi	1378	889
Meyerhof	1141	723
Brinch Hansen	1105	723
Vesic	1261	769
Eurocode 7	1230	748

Table 6.1. *Summary of values for bearing capacity (in kPa)*

6.3. Effects of soil variability on variability in bearing capacity and safety of the foundation

6.3.1. *Methodology*

Bearing capacity is calculated using the "Eurocode 7" model. We consider the random nature of the variation in properties, by assuming that the mathematical distribution of the friction angle and cohesion are known, and performing Monte Carlo simulations. A total of 3,000 numerical experiments are performed, from which the mean values, standard deviations and coefficients of variation of the results are all extracted. The friction angle ϕ' is assumed to follow a distribution such that tan ϕ' has a normal distribution, whose mean and standard deviation are μ[tan ϕ'] and σ[tan ϕ']. The cohesion C' is assumed to follow a log-normal distribution. We are giving μ[ln(C')] and σ[ln(C')]. The values for these parameters are listed in Table 6.2.

Configuration A $\phi' = 35°, C' = 0$ kPa	Configuration B $\phi' = 25°, C' = 20$ kPa
μ[tan ϕ'] = 0.70	μ [tan ϕ'] = 0.47
σ[tan ϕ'] = 0.05	σ[tan ϕ'] = 0.033
C' = 0	μ [ln(C')] = 2.75
	σ[ln(C')] = 0.70

Table 6.2. *Statistical properties of the mathematical distributions of the relevant random parameters*

Configuration A corresponds to a mean value of ϕ' that is close to 35°, with a standard deviation of ϕ' that is of the order of 2°. Configuration B corresponds to a mean value of ϕ' close to 25°, and a standard deviation of ϕ' of around 1.6° and a mean value of C' close to 20 kPa, with a standard deviation of the order of 16 kPa (the coefficient of variation is close to 100%, which is a consequence of the log-normal distribution of C'), which results in a few extremely large values being generated.

An additional degree of freedom for the simulation is the possibility of considering that the two properties of friction and cohesion could be independent or could be correlated.

A correlation could be present between the two quantities for two completely different reasons:

– a statistical reason: if the results of shear tests, performed on a particular soil sample, are used to identify C' and ϕ' by constructing the Mohr rupture diagram, there will be a specific uncertainty that results from variability within the samples and the uncertainties within the tests. An infinite number of lines are possible, and the estimates for C' and ϕ' are linked: if a smaller value is taken for the friction angle, this will result in a larger value for cohesion. This leads to an inverse correlation between these two properties;

– the second reason is a geotechnical one. If we now consider a set of samples obtained from sampling performed *in situ* in the soil, and we bear in mind the heterogeneity of the soil, this heterogeneity may be linked to a variable fraction F of clay particles. As F increases, the friction angle will tend to fall while the cohesion tends to increase. The result of this is that, for a heterogeneous soil sample, a dispersion between the two properties will be seen, with an inverse correlation.

In what follows, we assume that the negative correlation results from the geotechnical explanation. For configuration B, we will allow the degree of correlation $r\,(\phi',\,C')$ to vary, considering three different variants:

B1: $r = 0$ (independent parameters); B2: $r = -0.50$; B3: $r = -0.75$

The statistical distributions of C' are identical in all three cases, as can be seen from comparison of the cumulative distributions $F_{C'}(C')$ (Figure 6.3), but the correlation introduced between the two mechanical properties (Figure 6.4(a) and 6.4(b)) has important consequences for estimating the bearing capacity.

In fact, when a negative correlation is introduced, the largest values of ϕ' correspond to the smallest values of C'. The coefficients N_γ, N_q and N_c are monotonically increasing functions of ϕ' and the cohesion acts in the same manner on the bearing capacity. The existence of this negative correlation leads to a specific effect where the $C' N_c (\phi')$ term may decrease as ϕ' increases.

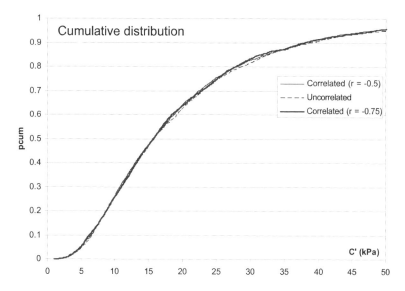

Figure 6.3. *Similarity between the cumulative distribution of C' for variants B1, B2 and B3, where pcum represents the cumulative distribution $F_{C'}(C')$*

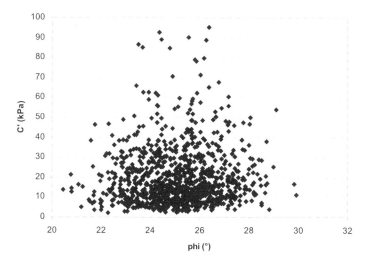

Figure 6.4a. *Illustration of the relationship between the ϕ' and C' parameters: B1 variant (no correlation, r = 0)*

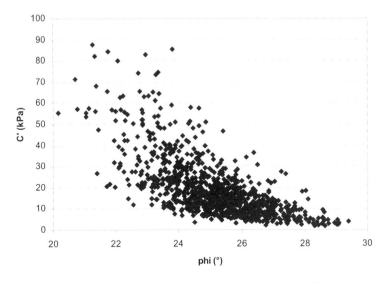

Figure 6.4b. *Illustration of the relationship between the ϕ' and C' parameters: B3 variant (negative correlation r = -0.75)*

6.3.2. *Purely frictional soil*

Figure 6.5 shows the results for two series of 3,000 simulations. The mean value is 1,273 kPa, and the standard deviation is 353 kPa, implying a coefficient of variation of 27.7%. The distribution is not, however, symmetric, which can be explained by the highly nonlinear nature of the mechanical model.

For the same reasons, the mean value of q_{tot} corresponds to a value of friction angle ϕ' that is slightly greater than the mean value of ϕ'. Strictly, this asymmetry formally precludes the use of the properties of the normal distribution to determine a given fractile using: $q_{tot\,k\,=\,5\%} = \mu\,(q_{tot\,k}) - 1.645\,\sigma\,(q_{tot\,k})$, but this would give a value close to 700 kPa.

If we instead identify the $q_{tot\,k\,=\,5\%}$ fractile from the distribution produced by the simulation, then we obtain a value of around 780 kPa (Figure 6.5).

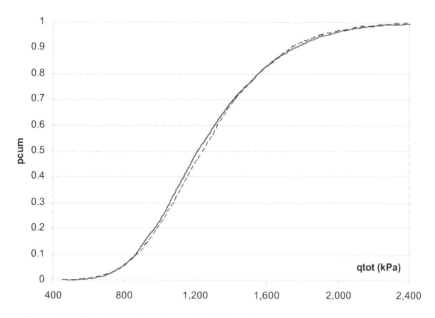

Figure 6.5. *Cumulative distribution for the load-bearing capacity, configuration A*

6.3.2.1. *Comparison with the semi-probabilistic approach from the Eurocode 7 regulations*

Eurocode 7 offers the designer a choice between several modeling approaches. Worthy of specific mention are [BRE 09]:

– approach 2, which involves applying a partial factor (PF) of 1.40 directly to the global value of the resistance output from the calculation, taking the characteristic values of the mechanical properties and the resistance model; and

– approach 3, which involves applying partial factors to the material properties, and then using these to determine the calculated value of the global resistance, by feeding these values into the mechanical model.

In our case, approach 2 involves determining ϕ'_k and then determining q_{tot} (ϕ'_k) using a Gaussian distribution of $\tan(\phi')$.

We have:

$$[\tan(\phi')]_{k=5\%} = \mu\,(\tan(\phi')) - 1.645\ sd\ (\tan(\phi')) = (0.7 - 1.645 \times 0.05) = 0.617$$

where sd is the standard deviation. The result is: $\phi'_k = 31.7°$ and $q_{tot}\,(\phi'_k) = 783$ kPa.

We then have:

$$q_{tot\ d} = q_{tot\ k\ =\ 5\%}/1.40 = 559\ kPa$$

Approach 3 uses the identified value of $\phi'_k = 31.7°$, so that, with PF (ϕ') = 1.25: $[\tan(\phi')]_d = [\tan(\phi')]_{k\ =\ 5\%}/1.25 = 0.4942$.

This results in $\phi'_d = 26.3°$. If we feed this value into the computational model, we obtain q_{tot} (ϕ'_d) = 392 kPa. This value is notably smaller than that obtained from the "global" calculation of Approach 2.

6.3.3. Soil with friction and cohesion

Figure 6.6(a) and (b) illustrate how the calculated bearing capacity correlates with the friction angle, for the cases with and without correlation. As might be anticipated in light of the results in Figure 6.4(a) and (b), the influence of the correlation between the two properties is significant.

In the absence of correlation, the monotonically increasing nature of the expressions that make up the mechanical model results in positive correlation between ϕ' and q_{tot}.

If we consider the case with correlation present (in this case, $r = -0.75$), the effect is counteracted by the greater probability of finding weak cohesions as ϕ' increases. That then results in an overall negative correlation between ϕ' and q_{tot} (Figure 6.6(b)). This modification is accompanied by a reduction in the variability of q_{tot}, due to the mutual counterbalancing of the effects of ϕ' with those of C'.

Table 6.3 summarizes some of these results.

	$r = 0$	$r = -0.5$	$r = -0.75$
μ (q_{tot}) (kPa)	765	756	747
σ (q_{tot}) (kPa)	348	293	249
cv (q_{tot})	0.455	0.389	0.334
μ (q_{tot}) – 1.645 σ(q_{tot}) (kPa)	192	274	337

Table 6.3. *Variation in statistical properties of the simulated carrying capacity (kPa) with correlation between cohesion and friction angle*

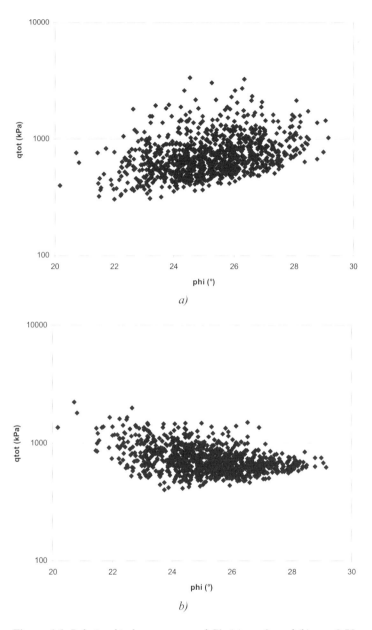

Figure 6.6. *Relationship between q_{tot} and C': (a) r = 0; and (b) r = -0.75*

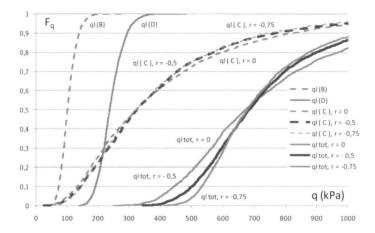

Figure 6.7. *Cumulative distribution F(q) for the bearing capacity (and contributing terms) with and without correlation between friction and cohesion*

Figure 6.7 plots the cumulative distributions of the three components of the bearing capacity (surface, depth and cohesion terms, written $ql(B)$, $ql(D)$ and $ql(C)$, respectively) and the total value q_{tot}, with and without correlation r. The first two terms $ql(B)$ and $ql(D)$ range from 0 to 200 kPa and are unchanged regardless of the degree of correlation that is present or otherwise. The overall distribution $ql(C)$ for the cohesion term remains almost unchanged, but conversely a difference appears in the q_{tot} term: when the correlation is considered, there is a slight reduction in the frequency of the smallest values, since the effect of a small value of ϕ' on the first two terms is compensated by a larger value for the cohesion term.

6.3.3.1. *Comparison with the Eurocode 7 semi-probabilistic approach*

In terms of approach 2 (global), due to the dependencies between variables it is no longer possible, as it was with configuration A, to find a simple expression for the characteristic value $q_{tot}(C'_k, \phi'_k)$. It is however possible to extract $q_{tot\,k}$ from the results of simulations, using the 5% fractiles of the simulated distributions (the equation $X_k = \mu(X) - 1.645\,\sigma(X)$ cannot be used, since that assumes a normal distribution). The design value is $q_{tot\,d} = q_{tot\,k}/1.40$.

We find:

$- r = 0$	$q_{tot\,k} = 404$ kPa	$q_{tot\,d} = 404/1.40 = 289$ kPa;
$- r = -0.50$	$q_{tot\,k} = 460$ kPa	$q_{tot\,d} = 460/1.40 = 329$ kPa;
$- r = -0.75$	$q_{tot\,k} = 508$ kPa	$q_{tot\,d} = 508/1.40 = 363$ kPa.

Approach 3 involves identifying the characteristic values and then the calculated values for the properties, and using these to determine the overall calculated value for the resistance. This approach does not allow us to take account of any correlation there may be between variables. The approach followed for configuration B gives:

$$[\tan(\phi')]_{k=5\%} = \mu\,(\tan(\phi')) - 1.645\,\sigma\,(\tan(\phi')) = (\,0.47 - 1.645 \times 0.033) = 0.415$$

which results in $\phi'_k = 22.6°$, such that, with PF $(\phi') = 1.25$;

$$[\tan(\phi')]_d = [\tan(\phi')]_{k=5\%}/1.25 = 0.333$$

which gives $\phi'_d = 18.4°$.

For the cohesion, we make use of the log-normal character of the distribution:

$$C_k = \exp\,(\mu\,(C') - 1.645\,\sigma\,(C')) = \exp\,(2.75 - 1.645 \times 0.7) = 5\ \text{kPa}$$

such that $C'_d = C'_k/1.25 = 4$ kPa.

It then follows that $q_{tot}\,(C'_d,\ \phi'_d) = 207$ kPa.

This value is (as for configuration A) much lower than that obtained using approach 2. It is also impossible to take account of correlations between variables, which is an important limitation of semi-probabilistic approaches, and one that in our case justifies the use of a probabilistic approach involving the use of Monte Carlo simulations.

6.4. Taking account of the structure of the spatial correlation and its influence on the safety of the foundation

6.4.1. *Spatial correlation and reduction in variance*

The Eurocode documents [CEN 07] introduce the concept of an "extended parameter" to take into account the fact that certain limit states are not defined in terms of the local value taken by a property at a given point, but rather by a representative value of this parameter over a volume of a particular size.

In order to identify the representative value, we must investigate both the spatial variability of the property and the manner in which this property affects the mechanical response of the system [BRE 09].

The properties of soils, as with many heterogeneous materials, can be described using regionalized variable theory [MAT 65]. The structure of the spatial correlation of these properties (friction angle, cohesion) is then represented by the choice of an autocorrelation distance and a correlation function.

In what follows, we will describe the structure of the spatial correlation of the friction angle, and we consider cohesion as a random variable that is correlated to the friction angle, using the same variance reduction technique. We will assume an autocorrelation function with an exponential form:

$$\rho(\tau) = \exp(-2\tau/l_c) \tag{6.3}$$

where l_c is the autocorrelation distance for the "friction angle" and "cohesion" properties (in order to keep things simple we will assume that both these properties have the same structure of spatial variability). This quantity is analogous to the range of the variogram discussed in section 5.4.1.

Note that the definition of the autocorrelation distance may vary significantly if an alternative correlation model is chosen [BAR 05]. For example, in the case of an exponential function, there is never, in a truly rigorous sense, an asymptotic limit, and a conventional definition of the range must be used.

It can be shown that there is a variance reduction function that corresponds to this autocorrelation function [TAN 84]:

$$\Gamma^2(L) = l_c^2/2L^2 \ (2\ L/l_c - 1 + \exp(-2L/l_c)) \tag{6.4}$$

This indicates that, if we assume the true variance of the property V_∞ to be known over a support of infinite extent, the variance observed over a finite support L will be smaller, and can be written in the form:

$$V_L = (1 - \Gamma^2(L))\ V_\infty \tag{6.5}$$

The complementary variance $V_\infty - V_L = \Gamma^2(L)\ V_\infty$ represents the *variance that will be observed between two samples of size L* (in our case, between two different foundations).

If, on the other hand, we assume that the variance between samples of size L_{labo} can be identified in the laboratory, then:

$$V_{exp} = V_\infty - V_{Llabo} = \Gamma^2(L_{labo})\ V_\infty \tag{6.6}$$

If we assume that a volume characterized by a dimension L is "affected" by movement in the foundation footing, the complementary variance can be expressed as:

$$V_\infty - V_L = \Gamma^2(L) \ V_\infty = V_{exp} \ \Gamma^2(L)/\Gamma^2(L_{labo}) \qquad [6.7]$$

This equation can be used to express the variance in terms of the experimental variance measured in the laboratory and the three dimensions l_c, L and L_{labo}.

We have already shown [BRE 05], [BRE 07] that it is in fact the ratios between these various dimensions that determines the degree of spatial variability.

The quantity L is not known explicitly. It represents the size of the volume of soil "governing the occurrence of a limit state" (as it is phrased in Eurocode). In the case of the bearing capacity of a foundation footing, it is logical to relate this volume to the size of the blocks involved in the limit analysis failure mechanisms. This size is linked to the length of the boundaries between blocks, and therefore to the dimensions B and D.

In what follows, we write $L = \alpha B$, where α is a scalar variable that takes the values 1, 3 and 5. Figure 6.8 illustrates the variance reduction function for 8 values of l_c lying between 1 m and 128 m (it is assumed that $L_{labo} = 0.15$ m).

As l_c tends to infinity (here 128 m), we recover the global variance. If, on the other hand, l_c is not infinitely large compared to B, we must take into accounts the effects of variance reduction. Specifically, this means that the variance associated with the "friction angle" parameter depends on the size of the volume being considered: averaging effects appear if a region of significant size is considered. The expression for the complementary variance can be used to quantify the amplitude of the variance that must be considered for an "extended" parameter, representing the "friction angle" property over a region of size L.

For the foundation problem, we therefore no longer need to consider the initial variance in the friction angle, but a reduced variance that takes account of this extended nature. The simulations must be supplied with the values of l_c and α. The result is a reduction in the variability of q_{tot}, which we have shown is directly linked to the variability of ϕ'.

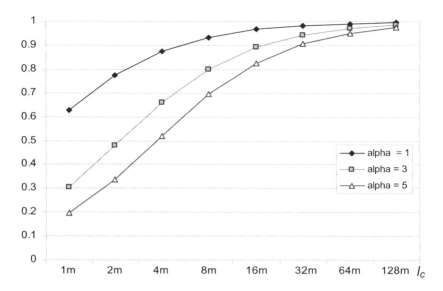

Figure 6.8. *Variance reduction $\Gamma^2(L=\alpha B)$ for eight different values of the correlation length l_c between 1 and 128 m*

6.4.2. *Taking account of the spatial correlation, and results*

The principles of these numerical simulations are exactly the same as in the previous cases. We vary: (a) the degree of correlation r; (b) the spatial correlation length l_c; and (c) the dimension α of the volume affected by a failure. We will only comment here on the results for configuration B, with those obtained for configuration A leading to similar conclusions. Figures 6.9 and 6.10 show the cumulative distribution for the bearing capacity for three different degrees of correlation r and for two different cases, without spatial correlation and with strong spatial correlation ($l_c = 2$m). When we take account of the spatial correlation, this significantly reduces the variance in the calculated bearing capacity, without modifying the mode of the distribution. Consequently, the values corresponding to outlying fractiles are significantly improved, as shown in Figure 6.10.

Figures 6.11 and 6.12 show, for $r = -0.75$, the variation in the standard deviation $\sigma(q_{tot})$ and the value $[\mu(q_{tot}) - 1.645\,\sigma(q_{tot})]$ for different values of l_c and α.

It can be observed that, as l_c tends to infinity, we recover the values given in Table 6.3 for the case of random variables (249 kPa for the standard deviation, 337 kPa for $[\mu(q_{tot}) - 1.645\,\sigma(q_{tot})]$). In this situation, the spatial variations in ϕ' are very slow: the values and distributions observed on the scale of laboratory samples can be used in calculations performed on the scale of the entire foundation footing.

This is no longer the case as the ratio $l_c/\alpha B$ decreases in value, either because l_c decreases or because α increases. In that case, the variability of q_{tot} is reduced and the value of $[\mu(q_{tot}) - 1.645\ \sigma(q_{tot})]$ steadily increases in a significant manner.

We note, however, that the computed values for $[\mu(q_{tot}) - 1.645\ \sigma(q_{tot})]$ do not represent fractiles in the rigorous sense of the term (we have a non-Gaussian distribution). More precise analysis of the fractiles can only be achieved by considering more detailed results from the simulations (or by assuming that the distribution is log-normal).

For example, in the case where $r = -0.75$ we find that the 5% fractile changes from 508 kPa without spatial correlation to 526 kPa if $l_c = 8$ m and to 554 kPa if $l_c = 2$ m, an increase of 10%.

When we take into account the spatial correlation in the soil properties (by reducing the variance to be considered in the calculation) this results in an improvement to the statistical distribution of the bearing capacity. If we do not take it into account, this represents a conservative approach that may lead to unnecessary overspending.

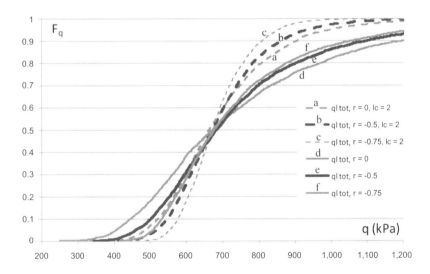

Figure 6.9. *Cumulative distributions F_q for the bearing capacity, with and without spatial correlation ($l_c=0$ m and $l_c=2$ m), for three levels of correlation ($r = 0, -0.5, -0.75$ m) between cohesion and friction*

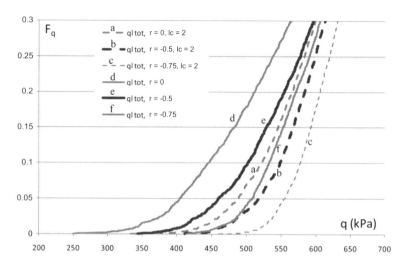

Figure 6.10. *Zoom view of the cumulative distributions F_q for the bearing capacity, with and without spatial correlation ($l_c=0$ m and $l_c=2$ m), for three levels of correlation ($r=0$, -0.5, -0.75 m) between cohesion and friction*

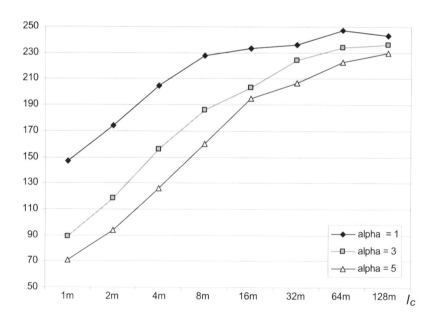

Figure 6.11. *Variation in standard deviation σ (q_{tot}) for various values of l_c and α*

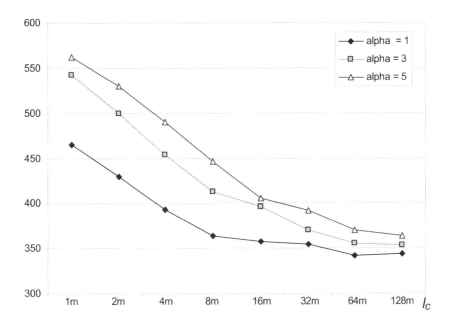

Figure 6.12. *Variation in [μ (q$_{tot}$) – 1.645 sd(q$_{tot}$)] for various values of l$_c$ and α*

6.5. Conclusions

6.5.1. *Conclusions drawn from case study*

Our analysis of the problem of the bearing capacity of a shallow foundation footing, including variability in spatial properties, leads to some important conclusions (but ones that are restricted to this example):

– the effects of model errors are negligible compared to errors in the variability of mechanical properties;

– a probabilistic approach (involving Monte Carlo simulations, since explicit solutions cannot be developed for this highly nonlinear problem) enables us to characterize the statistical distribution of the bearing capacity. It demonstrates the need to quantify in detail the variability of the mechanical properties of cohesion and friction angle;

– correct treatment of correlations between two parameters leads to very significant modifications to the result. This can only be achieved by building on statistically reliable raw data;

– the "Eurocode" semi-probabilistic approach yields results which are highly dependent on the approach used (the values obtained depend strongly on the values of the partial coefficients, which have been calibrated for the general case and are not necessarily appropriate for the specific case of interest). It is also incapable of taking account of correlations between parameters;

– correct treatment of spatial correlation also has a positive effect on the outlying fractiles for the bearing capacity, by reducing its variance. Not considering spatial correlation is a conservative approach, which may lead to unnecessary overspend. The main challenge is in identifying a realistic value for the correlation length. It should, however, be added that correct treatment of spatial correlation becomes indispensible when treating the questions involving differential settlement between neighboring foundation footings or when modeling soil-structure interactions, which is outside the scope of the present study [BRE 05], [BRE 07].

6.5.2. General conclusions

As was indicated in the introduction, our intention here was to illustrate the effects of taking into account material variability when considering the variability in the response of the mechanical system. To do this we considered the example of the bearing capacity of a shallow foundation footing. However, other problems can be treated using the same process: estimating the maximum bearing capacity of a reinforced concrete or wooden structure, estimating the lifetime of a reinforced concrete structure undergoing corrosion, evaluating the stability of a cut-off slope, etc. For every one of these problems:

– there is a need for a deterministic type of model (mechanical, chemical, etc.) that can describe the most important aspects of the phenomenon under study (resistance to flexion, penetration of aggressive agents, stability against slippage, etc.);

– there must be sufficient information available on the material properties to which the system in question is most sensitive: a statistical distribution but also any structure for the spatial correlation.

It is only when equal care is devoted both to the physical/mechanical description of the system and to the representation of material properties that it is possible to make a reasonable estimate of the reliability of the system under study. Of course, given the generally high cost of data acquisition (soil survey programs, laboratory tests, etc.), it is necessary to tailor efforts to real-world needs. Standards (such as Eurocode) are generally sufficient to treat contemporary problems, admittedly with many simplifications. Things are different when material variability is a key parameter determining the solution: problems of differential settlement, *in situ* evaluation of properties, etc. In such cases there is a need to use more sophisticated modeling approaches.

6.6. Bibliography

[BAR 05] BAROTH J., Analyse par EFS de la propagation d'incertitudes à travers un modèle mécanique non linéaire, Thesis, University of Clermont II, France, 2005, http://tel.archives-ouvertes.fr/tel-00397666/fr/.

[BRE 05] BREYSSE D., NIANDOU H., ELACHACHI S.M., HOUY L., "Generic approach of soil-structure interaction considering the effects of soil heterogeneity", *Symposium in Print "Risk and Variability in Geotechnical Engineering"*, *Geotechnique*, vol. 15(2), p. 143–150, 2005.

[BRE 07] BREYSSE D., LA BORDERIE C., ELACHACHI S.M., NIANDOU H., "Spatial variations in soil properties and their consequences on structural reliability", *Civil Engineering and Environmental Systems*, vol. 4(2), p. 73–83, 6/2007.

[BRE 09] BREYSSE D., *Maîtrise des risques et construction :Sécurité et réglementation*, Hermès, Paris, France, 2009.

[BRI 70] BRINCH HANSEN J., "A revised and extended formula for bearing capacity", *Danish Geotechnical Institute Bulletin*, no. 28, p. 5–11, 1970.

[CEN 07] CEN, Eurocode 7: Geotechnical Design - Part 1: General Rules, Brussels, 1996.

[MAG 04] MAGNAN J.-P., "Modèles de calcul des fondations superficielles: synthèse des méthodes", *Actes des conférences géotechniques (Param 2002 – Fondsup 2004 – ASEP-GI 2004)*, ENPC press, Paris, France, 2004.

[MAT 65] MATHERON G., *Les variables régionalisées et leur estimation*, Masson, Paris, France, 1965.

[MEY 63] MEYERHOF G.G., "Some recent researches on the bearing capacity of foundations", *Can. Geotech. J.*, no. 1, p. 16–26, 1963.

[ORR 08] ORR T.L., BREYSSE D., "Eurocode 7 and reliability based design", *Reliability Based Design in Geotechnical Engineering: Computations and Applications*, p. 298–343, Ko-Kwang Phoon, Taylor & Francis, London, 2008.

[TAN 84] TANG W.H., "Principles of probabilistic characterization of soil properties", in *Probabilistic Characterization of Soil Properties, Bridge Between Theory and Practice*, ASCE, Bowles & Co, Reading, 1984.

[TER 43] TERZAGHI K.V., *Theoretical Soil Mechanics*, p. 423–427, John Wiley & Sons, Chichester, 1943.

[VES 73] VESIĆ A.S., "Analysis of ultimate loads of shallow foundations", *Journal of Soil Mechanics and Foundations Division*, ASCE, vol. 99(1), p. 113–125, 1973.

PART 3

Metamodels for Structural Reliability

Introduction to Part 3

Part 3 considers a class of reliability models known as "metamodels", or response surfaces. This application of response surfaces to the problem of structural reliability has its origin in the need to represent the response of a system (a mechanical system but also a living biological system, a social system, or an economic system) under the influence of stimuli. These stimuli are the input parameters to the system, and they may be poorly known or uncertain. The response surface is not therefore an exact representation of the response of the system, but rather an approximation – the quality and range of validity which we would like to evaluate. The main motivation in structural reliability is, in the case of analytical (and hence explicit) response surfaces, to facilitate the coupling of probabilistic algorithms with the model (assumed exact) that determines the response of the system.

This part is divided into two chapters. Chapter 7 introduces physical and polynomial response surfaces, along with reliability calculations as applied to those surfaces. Chapter 8 discusses reliability calculations that use polynomial chaos expansion of the response surfaces. A range of analytical examples are presented. A trellis beam and a building framework are used to illustrate the concepts presented in these two chapters.

Chapter 7

Physical and
Polynomial Response Surfaces

7.1. Introduction

Generally, structural reliability analysis is based on the supply of mechanical and probabilistic models and a limit state function. In this chapter, we first define a mechanical model that describes structural behavior. In a general sense, the mathematical transfer function M allows us to evaluate the influence of loading with the knowledge of input parameters (or stimuli) that describe the structure and its environment. These parameters constitute the vector x. The model response is denoted here as $y = \mathcal{M}(x)$.

Next, we define a probabilistic model for the input parameters that are considered to be poorly known or uncertain, even from a statistical analysis of data samples when they are available, or by expert judgment and a database [JCS 02]. This probabilistic model is characterized by the joint density of the input random variable X, denoted $f_X(x)$.

Thirdly, and finally, we define a limit state function that mathematically translates the failure criterion against which the structure must be justified. This function is written with the general form $g(\mathcal{M}(X), X')$ and is based on the effect of loading to which we fix limits, gathered in a vector X'.

Chapter written by Frédéric DUPRAT, Franck SCHOEFS and Bruno SUDRET.

By denoting $f_{X,X'}$ the joint density of vectors X and X', the objective of the analysis is to evaluate the structural reliability through one estimate such as the probability of failure P_f defined by:

$$P_f = \int_{\{x:g(\mathcal{M}(x),x')\le 0\}} f_{X,X'}(x,x')\,dx \qquad\qquad [7.1]$$

Several methods exist for solving the problem [DIT 96], [LEM 09]. Among them, in this chapter, only a Monte Carlo simulation and FORM method are presented and used.

7.2. Background to the response surface method

Many scientific fields and trends of thought have contributed to the elaboration of the so-called Response Surface Methodology (RSM). The beginnings of the response surface approach appeared a few years before Box & Wilson's developments [BOX 50]. Various scientific fields were involved:

– animal and vegetal biology, and the building of growth curves [REE 29], [WIN 32], [WIS 39];

– human sciences and the analysis of the response of a population to stimuli [BLI 35a], [BLI 35b], [GAD 33]; these works were based on those of the psychiatrist Fechner in 1860;

– agronomy and the study of soil fertilization [CRO 41], [MIT 30], [STE 51].

These approaches were based mainly on the basic assumption (mathematically justified by the Weierstrass's theorem) that, under some conditions of regularity, a response can be represented by polynomials. Thus, within this context, in 1951 the chemists Box and Wilson developed the concept of a response surface, relying both on analytical regression techniques and the building of experiences. In particular, they had already carefully described the need to pay attention to the choice of stimuli variables, and to the allocation of their relative weight. The empirical models they developed have been enriched by the definition of observation periods [BOX 55] and by error computation [BOX 57]. With the increase of the number of potential models, selection criteria were provided [BOX 59a] such as the generalized variance minimization of variable estimation [BOX 59b]. The period 1950–1970 was a fruitful one with the appearance of three major scientific developments with probabilistic insights:

– research into optimal functional representation through stochastic approximation in the presence of outliers; an expansion to multivariate problems was proposed [KIE 52], [ROB 51];

– the comparison of growth curves in biometry [ELS 62], [RAO 58], where response functions come from projections on a family of orthogonal polynomials. Their coefficients are then used for forecast studies;

– the theory of optimal models under constraint in the case of linear models [KIE 60]. The optimization of a response function was linked to the minimization of the generalized variance of parameters.

This last trend of thought offers a well-structured theoretical contribution that was soon used as a reference by others. Numerous works about optimization appeared after it [ANS 63], [NEL 65], [POW 65]. The end of this period saw the emergence of non linear models. The most significant works were certainly carried out on inverse polynomials [NEL 66] and the statistical estimation of parameters from experimental processes [ATK 68]. Nevertheless, the increase in use of nonlinear models was really significant, due to the increase in computational capacity of computers. The criteria for model validation appear to be very specific to each application field; review papers in biometry [MEA 75] and in the nuclear field [HEL 93] can be cited as examples.

7.3. Concept of a response surface

7.3.1. *Basic definitions*

The term "response surface" denotes the wish to develop a formal representation based on geometrical ideas; it is surface building in the probabilistic space of the response of a physical process to stimuli. The property being studied, or response Y, is the result of a transfer function that characterizes the sensitivity of a system to input parameters. This response then varies with the variation of input parameters known as *stimuli*. These are modeled by random fields or variables, denoted X_i, $i=1,..,n$, and then characterized by a set of available statistical information, denoted θ_j, $j=1,..,$ p, (independent or correlated probability density functions, normalized moments, etc.). These random variables (or fields) are called *basic random variables* (or fields). This transfer is represented in Figure 7.1.

Figure 7.1. *Response of a transfer function to stimuli, modeled by random fields or variables*

The modeling of random fields is sometimes needed for the representation of spatio–temporal variations of uncertain input parameters. Modeling with random variables, which is simpler, is nevertheless sufficient for specific problems. For simplicity in the following sections, we use random variables to represent input parameters.

Generally, knowledge of a transfer function as an explicit form of basic random variables is not available. We therefore look for an approximation, called a response function, M, which is often selected from amongst a family of usual functions, linear or not, characterized by random or deterministic parameters χ_k, $k=1,...,l$. These parameters are deduced from the fitting of the response to the experimental data. The geometrical representation, with a curve, a surface, or a hypersurface, is called a response surface. The introduction of geometrical tools such as contour lines onto this response surface can then be used as frontiers of the safety domain. To build a response surface, we must provide:

– $X=\{X_1..., X_n\}$, a ranked set of representative random variables;

– $\theta=\{\theta_1...,\theta_p\}$, a set of statistical information about X (independent or correlated probability density functions, normalized moments, etc.);

– $M(X/\theta)$, an approximation of the response Y, formulated as an explicit function of X knowing θ, and obtained by the fitting of the set of parameters χ;

– $|.|$, a metric in the probabilistic space of basic random variables and responses.

The quality of fit of the approximation M to the response Y is then measured.

The response function can then be formally written as in Figure 7.2.

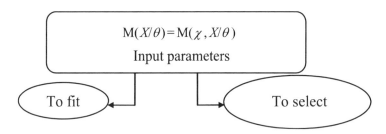

Figure 7.2. *Formal writing of a response function*

7.3.2. *Various formulations*

The choice of the type of formulation for the response function is made on the basis of specific criteria coming from the selected scientific methodology for studying the phenomenon. The first questions are the conceivable level of complexity, the availability of a complementary experimental approach and the actual state of knowledge. Two methods are possible but the building of a response surface is more and more frequently based on a mixed solution:

– the matching of an approximated model of the transfer by using usual mathematical functions (especially polynomials) and a selected database. [SCH 96] presents and compares the usual models used;

– the use of underlying deterministic physical laws in which random variables are introduced to account for intrinsic variations (height and period of a wave, speed of the wind for instance) or for uncertainties on modeling parameters [LAB 96]. Two issues then govern the building of the response surface, the physical meaning of the deterministic models and the selection of the random variables, i.e. those that govern the variations of the studied quantities.

Finally, the difficulty of statistically characterizing the basic variables adds to the difficulty of selecting the analytical formulation of the transfer. In the case of response surfaces that describe limit states, this question conditions the reliability measure. For instance, in the field of unidirectional laminated composite materials loaded in the direction or orthogonal to the direction of fibers, we consider generally that a limit state can be deduced from three criteria: stresses, external loading and the geometrical size of the material. Various assumptions on the number of random variables and the typology of their distribution can lead to variations of more than 30% of the value of corresponding safety factors [NAK 95].

7.3.3. *Building criteria*

Building criteria are specific to each application field. Thus, in the following discussion, criteria are ranked according to their importance for problems relative to the safety of buildings and structures. These criteria nevertheless raise questions that can be extended to other fields. Further, it would be wrong to consider that a unique solution exists. The final choice is the result of an optimization under constraints that we have proposed.

We aim especially to take benefit of the increased power and computational capacities of computers which offer the ability to refine the mathematical representation and control the intrinsic uncertainties due to model fitting. However, one must always keep in mind the requirement of the physical meaning. Thus, the major elements in our approach are:

– the physical meaning of the representation;

– the effects of the choice of probabilistic modeling;

– the measure of the quality of the fit;

– a reduction of the level of complexity, consistent with acceptable computational costs.

7.3.3.1. *Physical meaning of the representation*

An understanding of the physical mechanism that underlies the physical phenomenon is fundamental when choosing the set of input variables and the approximation function. This criterion can lead to the necessity to base the formulation of the function on deterministic relationships. Intrinsic randomness is then introduced through random variables and we account for the model uncertainty through random parameters. The use of deterministic relationships and a careful selection of basic variables, if an analytical relationship is available, are shown to be more realistic than the use of fitted models with the usual mathematical formulations.

7.3.3.2. *Effects of the choice of probabilistic modeling*

The probability law of the system response depends on the probabilistic characterization of the input parameters (probability distribution, scatter, skewness, kurtosis, etc.). In a simple case where the transfer function M is a linear function of normally distributed random variables, the response is normally distributed too. In the general case, random variables are non-normally distributed and the transfer function is more or less nonlinear.

To control the effect of this degree of nonlinearity of the transfer function, several works suggest approaching this function with a polynomial approximation (generally linear, quadratics or cubic). A linear approximation is usually not sufficient [BOU 95]; cubic or fifth order approximations allow us to assess in some cases the moments of third and forth order in a satisfactory way. Polynomial approximations can be of high order and costly in terms of identification of their parameters.

Actually, to guarantee the transfer of distribution laws, it is necessary to control the good fit of the Jacobian matrix $[D(X)/D(Y)]$ [LAB 95], [SCH 08]. We then consider a response surface with a single variable of the form $Y = M(X)$ with M, a bijective monotonic function which can be derived, and X a basic variable. Knowing the probability density f_X of X, we can compute f_Y, the probability density of Y. G is the cumulative function associated with the probability density function g, and we know that:

$$GY(y) = P(Y < y) \text{ and } fY(y) = \frac{d}{dy} P(Y < y)$$

$$f_Y(y) = P[X < M^{-1}(y)] \text{ and finally} \qquad f_Y(y) = f_X[M^{-1}(y)]\left|\frac{1}{M'}[M^{-1}(y)]\right|$$

where $|.|$ denotes the absolute magnitude.

In the more general case of a multivariate problem for which X and Y are vectors, probability density functions of random input and output are linked by the relationship:

$$f_Y(y) = f_X(x)[M^{-1}(y)]\left|\frac{D(X)}{D(Y)}\right|$$

Thus, to obtain a good approximation of P_Y, a good fit of function M should be reached as well as a fitting of the partial derivatives. The linear, quadratic or cubic polynomial functions described before have, respectively, constant, linear and quadratic derivatives. The difference between these three functions is generally significant near the bounds of the studied domain (realizations of basic variables), and the perturbations on distribution tails can be significant and thus modify the results of reliability computation. Thus, the choice of low order polynomial, very convenient from a computational point of view, can lead to false probability functions for the response even they seem to correctly represent the trend. Such a choice is very sensitive to the effects of the choice of probabilistic modeling of input parameters.

7.3.3.3. Measurement of the quality of fit

We aim here to define a metric (a measuring tool) that gives a rational tool to quantify the quality of fit. Usual metrics, known as second order metrics, allow us to obtain the variables that are dominant in the response because we can quantify their influence on the variance of the response. They are thus not very effective when singular events external to the distribution functions of the input variables occur, and they only give an indication of the measure of uncertainty [BIE 83], [IMA 87]. To solve these shortcomings, metrics based on inter-quantile discrepancies [KHA 89] or on the measurement of the system entropy [PAR 94] are available.

Another approach consists of the use of regression metrics. Let us denote the response function for which we want to fit the parameters as f, and the error as ε. Let us consider, for example, a regression model such as:

$$Y = M(X/\theta) + \varepsilon$$

When the error ε is supposed to be normally distributed with 0 mean and a diagonal covariance matrix, then the fitting of f with least square and maximum likelihood methods are identical. The L_2 metric (integration of the square of the residual u) is then the most efficient. The L_1 metric (integration of the norm of the residual) is more efficient for an exponential distribution of the error, which is then more scattered.

It has been shown in the previous section that the fitting of a Jacobian matrix of partial derivatives is very interesting. Thus, it also seems very interesting to choose a metric already available in variational theory:

$$\|u\| = \sqrt{\|u\|_{L_2}^2 + \sum_{i=1,\ldots,n} \left\| \frac{\partial u}{\partial X_i} \right\|_{L_2}^2}$$

with $\|\ldots\|_{L_2}$, L_2 norm in the Sobolev space H_1 :

$$H_1 = \left\{ u \left| \frac{\partial^\alpha u}{\partial X_i^\alpha} \in L^2 \right. \text{ (second order integrable) for } \alpha = 0, 1 \text{ and } i = 1\ldots,n \right\}$$

The underlying idea of this choice of metric is thus to prefer the better control of the distributions tails through successive transfers, in comparison to the control of the central part, by the fitting of the first moments. Every building of a response surface should be suggested with a metric that conditions the sense of the approximation and allows us to explain some limits in the representation.

7.3.3.4. Reduction of complexity level and tractability for computations

To gain accuracy, models of a high order could be interesting. This increase in the computational procedures (optimization algorithms for the fitting under constraint) must, however, be justified: this increase in the complexity level leads to an increase of the computational costs that should be kept as reasonable as possible.

For more details, this question is illustrated in [SCH 07] through several studies concerning wave–structure interaction, which look at the effects of the order of the Stokes kinematics model, of accounting for the inertia term in load computing, and the number of elements needed for the integration of distributed loading on the beam.

7.4. Usual reliability methods

7.4.1. *Reliability issues and Monte Carlo simulation*

Structural reliability aims to assess the failure probability of a mechanical system in a probabilistic framework according to a failure scenario. Structural reliability methods allow not only the probability of failure P_f to be computed for a single component of a system, or for a system as a whole then involving several interacting components, but also the sensitivity of this probability against each random variable of the problem to be determined [DIT 96], [LEM 09].

A failure criterion can be expressed thanks to a performance function $g : \mathbb{R}^M \to \mathbb{R}$ conventionally described by parameters in such a way that $D_f = \{x : g(x) \leq 0\}$ defines the *failure domain* and $D_s = \{x : g(x) > 0\}$ defines the *safety domain*. The frontier $\partial D = \{x : g(x) = 0\}$ is the *limit state surface*. The probability of failure is:

$$P_f = \mathbb{P}(g(X) \leq 0) = \int_{D_f} f_X(x) \, dx \qquad [7.2]$$

where f_X is the probability density function of X.

When the performance function is analytically expressed, for instance when a response surface is implicated in the expression of g, then the integral over the implicitly defined integration domain is numerically feasible from Monte Carlo simulation: N_{sim} simulations of the input vector X are supplied, and for each of them g is evaluated. If N_f is the number of simulations for which g is negative, the probability of failure P_f is approximated by the ratio N_f / N_{sim}. The method is fairly simple but computationally very costly: about $N_{sim} \approx 4.10^{k+2}$ samples are required if a 5% accuracy is expected when P_f is about 10^{-k}. In usual practical applications, k lies from 2 to 6, which compromises the use of the method, except if a consistent response surface is available. Among methods which have been developed to circumvent this drawback, the First Order Reliability Method (FORM) is one of the most employed.

7.4.2. *FORM*

FORM provides an approximate value of the failure probability by recasting the problem in a *reduced centered Gaussian space*, where all random variables ξ are

Gaussian with zero mean and unit standard deviation. An iso-probabilistic transformation T is needed for this goal $T : X \to \xi(X)$. For independent random variables with marginal cumulative probability function $F_{X_i}(x_i)$, this transformation simply reads $\xi_i = \Phi^{-1}\left(F_{X_i}(x_i)\right)$ where Φ is the cumulative reduced centered Gaussian probability function. For correlated random variables, Nataf or Rosenblatt transformations can be resorted to [LEM 09], (Chapter 4). Equation [7.2] then becomes:

$$P_f = \int_{\{\xi\,:\,G(\xi)\equiv g(T^{-1}(\xi))\leq 0\}} \varphi_M(\xi)\,d\xi_1 \ldots d\xi_M \tag{7.3}$$

where $G(\xi) = g\left(T^{-1}(\xi)\right)$ is the performance function in the reduced space and φ_M is the reduced centered multinormal probability density function of dimension n, defined by $\varphi_M(x) = (2\pi)^{-\frac{M}{2}} \exp\left[-\frac{1}{2}\left(x_1^2 + \ldots + x_M^2\right)\right]$.

A maximum value of φ_M is encountered at the origin of the reduced space, and φ_M exponentially decreases with respect to the distance $\left\|\xi^2\right\|$ from the origin, all the more since the number of variables is high. In the failure domain, the points contributing most to the integral [7.3] are therefore those which are closest to the origin. The second stage of FORM is to determine the so-called design point P^* which is the point of D_f closest to the origin, and so the most probable failure point. This point is the solution of the optimization problem:

$$P^* = \underset{\xi \in \mathbb{R}^M}{\text{Arg min}}\left\{\frac{1}{2}\|\xi\|^2 \;/\; G(\xi) \leq 0\right\} \tag{7.4}$$

A suitable optimization under constraint algorithm can solve [7.4]. The reliability index β is then defined as the algebraic distance from the origin to the limit state surface ∂D : $\beta = \text{sign}\left(G(0)\right)\left\|\xi^*\right\|$. Once ξ^* is determined, the limit state surface ∂D is approximated by a hyper-plan tangential at P^*. The integral [7.3] is then reduced to $P_f \approx P_{f,FORM} = \Phi(-\beta)$. The linearized limit state is expressed by $\tilde{G}(\xi) = \beta - \boldsymbol{\alpha}^{\mathrm{T}} \cdot \xi$. The unit vector of direction, cosine $\boldsymbol{\alpha}$, which is perpendicular to the hyper-plan, allows the sensitivity factors to be computed by α_i^2 for each independent random variable X_i.

FORM is fruitful because it leads to a fairly good approximate value of P_f with a reasonable computational cost. The approximation is better when β is large, and is often relevant and satisfactory provided that the design point P^* is consistent (i.e. with no local *minima*). Moreover, sensitivity measures and importance factors are given which are matters of interest for the designer.

Readers are invited to refer to [LEM 09] to investigate SORM (Second Order Reliability Method), based on a quadratic approximation of the limit state function. SORM is *a priori* more costly than FORM, but also more accurate.

7.5. Polynomial response surfaces

Amongst other possible forms of response surfaces (RS), polynomial response surfaces have been widely used for reliability problems in mechanics. In the following section, only simple polynomial response surfaces are addressed, in contrast to those based on polynomial chaos in Chapter 8.

7.5.1. *Basic formulation*

If $X=\{X_i, i=1,\ldots,M\}$ is the random variables vector, the quadratic response surface $\hat{g}(X)$ of the true limit state function $g(X)$ is expressed by:

$$\hat{g}(X) = a_0 + \sum_{i=1}^{M} a_i X_i + \sum_{i=1}^{M} a_{ii} X_i^2 + \sum_{i=1}^{M} \sum_{j=1 \neq i}^{M} a_{ij} X_i X_j \qquad [7.5]$$

where $a=\{a_0,a_i,a_{ii},a_{ij}\}^T$ is the vector of unknown coefficients. These coefficients are obtained by least square method from the Numerical Experimental Design (NED) $\{x^{(k)}, k=1,\ldots,N\}$, where the number of sampling points N is at least equal to the dimension of a:

$$a = \underset{a}{\arg\min} \sum_{k=1}^{N} \left(y_k - \hat{g}\left(x^{(k)}\right)\right)^2 \qquad [7.6]$$

where $y_k = g\left(x^{(k)}\right)$ is the value of the true limit state function at point $x^{(k)}$. From $\hat{g}(X) = \left\{1, X_i, X_i^2, X_i X_j\right\}\left\{a_0, a_i, a_{ii}, a_{ij}\right\}^T = B(X)^T a$, a is computed by [FAV 89] as:

$$a = \left(C^T C\right)^{-1} C^T y \qquad [7.7]$$

where C is the matrix whose rows are the vectors $B\left(x^{(k)}\right)^{\mathrm{T}}$ and y is the vector of components y_k.

The basic formulation is simple. Nevertheless, some difficulties and controversial discussions arise when we consider the following points:

– working space. In order to assure the consistency of the NED, in such a way that the mechanical model behaves well, a physical space is preferable. Nonetheless, building the NED in a standardized space is easier. Indeed such a space is non-dimensional and allows the distance between sampling points to be efficiently controlled, which is useful to avoid the ill-conditioning of C. Moreover, it is a natural space for determining the reliability index;

– number of sampling points. When increasing the number of sampling points the least squares regression is improved (F-statistics and variance of the unknown coefficients a) but not necessarily the quality of the approximation (adjusted R^2 or crossed Q^2 correlation coefficients), which is better if the number of sampling points is just equal to the number of unknown coefficients. Outside of the NED, the relevance of the RS (Response Surface) is improved when the number of sampling points is higher, but if reducing the computational cost is an aim (compared to the cost obtained when the true failure function is used), then the number of sampling points should be minimized;

– topology of the NED. When the NED is compact, the quality of the approximation is improved but the domain of suitability is reduced. The location of the sampling points should depend on the behavior (the sensitivity) of the true limit state function;

– validity of the RS. Depending on what is being sought (reliability index and/or assessment of the failure probability) the validity of the RS should be estimated locally or globally;

– adaptability of the RS. It is often necessary to rebuild the NED because the domain of final utilization of the RS (the region of the most probable failure point) is far from the mean point (in case of low failure probability), which usually plays the role of initial central point of the sampling grid. A sequential procedure is therefore needed to rebuild the NED with respect to criteria for the consistency of the RS and in conjunction with probabilistic results or procedures;

– the order of the polynomial should be less than or equal to the unknown degree of nonlinearity of the true failure function, in order to facilitate the solving of the linear system;

– the presence of mixed terms in the expression of the RS contributes to capturing the effect of interaction between variables.

7.5.2. *Working space*

In the framework of structural reliability two working spaces are possible: the physical space *X* (dimensional random variables with any distribution and possible correlation) and the standardized space *U* (reduced centered Gaussian and uncorrelated random variables). Both spaces are simultaneously required to compute the unknown coefficients only if the RS is built in a standardized space (i.e. if the mechanical model is indeed defined in a physical space). The transformation from *X*-space to *U*-space is nonlinear, and modifies the topology of the NED and the failure surface as well [DEV 97] (see Figure 7.3 in the case of a factorial design). Numerous authors have hence chosen to operate in a physical space [BUC 90], [GAV 08], [KAY 04], [KIM 97], [MUZ 93], [RAJ 93]. If the searching procedure for the reliability index is associated with the sequential procedure for building the NED, it is, however, clearly preferable to operate in a standardized space [DEV 97], [DUP 06], [ENE 94], [GAY 03], [GUP 04], [NGU 09].

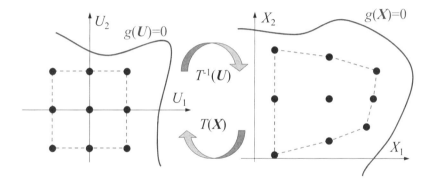

Figure 7.3. *Effect of the space shift*

7.5.3. *Response surface expression*

Whatever the working space chosen, the objective is to provide a satisfactory estimate of the reliability and, if possible, with a lower computational cost than that resulting from the use of the true limit state function. To do this, if second order estimates of the reliability are sought, linear and quadratic polynomial RS are good candidates. Respectively, the expressions of the RS are:

$$
\begin{aligned}
\hat{g}_I(X) &= a_0 + \sum_{i=1}^{M} a_i X_i \\
\hat{g}_{II}(X) &= a_0 + \sum_{i=1}^{M} a_i X_i + \sum_{i=1}^{M} a_{ii} X_i^2 + \sum_{i=1}^{M} \sum_{j=1 \neq i}^{M} a_{ij} X_i X_j
\end{aligned}
\qquad [7.8]
$$

Nevertheless if Monte Carlo simulations are expected to be used to assess the failure probability in a more or less large zone around the most probable failure point, or if the degree of nonlinearity of the failure function is from evidence higher than 2, increasing the order of the polynomial could be required [GAV 08], [GUP 04]. One method proposed to select the best-fitted order of polynomial involves the use of Chebyshev polynomials, a statistical analysis of the Chebyshev polynomial coefficients, and a statistical analysis of high-order RS [GAV 08]. The increase in accuracy brought by high-order RS is, however, counteracted by the added computational cost of the selecting procedure.

7.5.4. *Building the numerical experimental design*

7.5.4.1. *Number and layout of the sampling points*

The minimum number of points is the number of unknown coefficients, namely $N = (M+1)$ for a linear RS, $N = (2M+1)$ for a quadratic RS without mixed terms and $N = (M+1)(M+2)/2$ for a quadratic RS with mixed terms. With such a minimum number of points, only an interpolation is provided and the statistical significance of the coefficients is very poor. On the other hand, the use of the RS beyond the frontier of the NED is not expected to be consistent. Even if saving runs of the true failure function is one of the objectives, it is not desirable to limit the number of points to the minimum, due to the resulting poor quality and suitability of the RS.

A uniform layout of the sampling points around a central point is commonly adopted at the initial stage of building (central, central composite or factorial design). The number of sampling points of the NED depicted in Figure 7.3 is equal to $(2M+1+2^M)$ or (3^M) depending on whether one or more points are located out of the axes. Along the axis i the distance between points is stated as a function of the standard deviation of the variable:

$$x_i^{(k)} = X_{C,i} \pm h_i \sigma_i \tag{7.9}$$

where $x^{(k)}$ is the kth point of the NED and X_C is the central point. Similarly, in a standardized space:

$$u_i^{(k)} = U_{C,i} \pm h_i \tag{7.10}$$

Except in the case of adaptive procedures, the factor h_i is constant ($h_i = h$).

Too large or too small values of h compromise the quality of the RS and lead to erroneous results. An example is given in [GUA 01] where 21 random variables are involved (for the reliability of a portal frame) and a quadratic RS without mixed terms is employed: the failure probability shifts from 10^{-4} to 0.94 when h varies globally from 0.1 to 5. When h varies locally from 1.8 to 2.2, the failure probability and the reliability index, respectively, shift from 9.3×10^{-5} to 0.37 and from 0.06 to 4.47.

In order to avoid the ill-conditioning of the system C [1.7] the value of h must be large enough. Values of h lying in the range 1 to 3 are often suggested [BUC 90], [DEV 97], [DUP 06], [ENE 94], [KAY 04], [KIM 97], [MUZ 93], [NGU 09], [WON 05]. Minimum values of h have been also proposed in the range 0.2 to 0.5 [DUP 06], [ENE 94].

7.5.4.2. *Adaptive procedures*

The main goal of adaptive procedures is to combine a satisfactory approximation of the true failure function, at least in the neighborhood of the most probable failure point, with a limited computational cost. The following items are part of the adaptation:

– shape of the RS. A first linear ($\hat{g}_I(X)$) or quadratic without mixed terms ($\hat{g}_{II}(X)$ with $a_{ij}=0$) RS is attractive due to the small number of unknown coefficients and the computational cost needed to determine the most probable failure point. A refinement can be then undertaken with a quadratic with mixed terms RS [GAY 03], [NGU 09];

– initial NED. If the role played by the random variables is known *a priori* (for instance from the engineer's knowledge), a first central point X_{C1} can be located at $X_{C1,i} = \mu_i - h_i \sigma_i$ if X_i contributes to the non-failure and at $X_{C1,i} = \mu_i + h_i \sigma_i$ if X_i contributes to the failure. When no information is available on the role of variables, a pre-selecting procedure can also be employed with a few runs of the true limit state function [GAY 03];

– second central point. This can be located with respect to the first most probable failure point thanks to an interpolation [BUC 90] or found at the same location provided that the latter is situated inside the first NED [DEV 97], [ENE 94];

– mesh of the sampling grid. The size of the mesh is generally reduced as iterations proceed, in order to concentrate the NED in the region of the most probable failure point [DEV 97], [ENE 94], [GAY 03], [KAY 04], [MUZ 93]. Some considerations can be added in order to refine the mesh according to the number of random variables [KIM 97], the sensitivity of the RS towards the variables [DUP 06], [NGU 09], or specific statistics (confidence interval on the coordinates of the most probable failure point) [GAY 03];

– weighting of the sampling points. The unknown coefficients are computed, introducing a weighting diagonal matrix in equation [7.7]. The weighting factors depend on the closeness of points to the failure surface [KAY 04] or to the last most probable failure point [NGU 09];

– update of the NED. In order to limit the computational cost, it is of interest to keep in the NED all the points where the value of the true limit state function has already been computed. However, some of them, that could harm the quality of the RS, must be excluded according to suitable criteria [DEV 97], [DUP 06], [ENE 94], [GAY 03], [NGU 09].

7.5.4.3. Quality of the approximation

In order to verify the quality of the response surface, a classic measure of the correlation between the approximate and the exact value of the limit state function is the adjusted correlation factor:

$$R^2 = 1 - \frac{(N-1)\sum\limits_{k=1}^{N}\left(\hat{g}\left(x^{(k)}\right) - y_k\right)^2}{(N-N_c-1)\sum\limits_{k=1}^{N}\left(y_{mean} - y_k\right)^2} \qquad [7.11]$$

where y_{mean} is the mean value of the limit state function over the NED. If R^2 is less than 0.9, the quality of the response surface has to be improved. The cross correlation factor Q^2 is also employed (see Chapter 8).

7.5.5. Example of an adaptive RS method

7.5.5.1. General description [NGU 09]

7.5.5.1.1. First iteration

The NED is centered at U_C and comprises U_C and one point along each axis located at:

$$u_i^{(k)} = U_{C,i} + h_i \quad \text{with} \quad h_i = -\frac{h_0}{\left\|\nabla\hat{g}(U_C)\right\|}\frac{\partial\hat{g}(U_C)}{\partial U_i} \qquad [7.12]$$

where $u^{(k)}$ is the kth point of the NED and h_0 lies basically between 1 and 3. For the first iteration, the RS is linear in the standardized space $\hat{g}(U) = \hat{g}_1(U)$, the central point is the origin of the space $U_{C1} = U_0$ and a fictitious gradient of $\hat{g}(U_{C1})$ is considered, based on engineering knowledge in such a way that $h_i = \pm 1$.

According to [KAY 04], the weighting diagonal matrix W is introduced in equation [7.7] which becomes:

$$a = \left(C^T W C\right)^{-1} C^T W \, y \qquad [7.13]$$

where the weighting factors are expressed by

$$W_k = exp\left(-\frac{g\left(u^{(k)}\right) - y_{min}}{g(U_0)}\right) \qquad [7.14]$$

In [7.14], y_{min} is the minimum value of the true limit state function over the NED. Once the unknown coefficients a of $\hat{g}(U)$ are computed by [7.13] the first most probable failure point $U^{*(1)}$ is determined by FORM.

7.5.5.1.2. Second iteration

For the second and further iterations, the RS has a quadratic form $\hat{g}(U) = \hat{g}_{II}(U)$ with $(M+1)(M+2)/2$ being unknown coefficients. As suggested in [BUC 90], the central point of the second NED is stated by:

$$U_{C2} = U_0 + \left(U^{*(1)} - U_0\right)\frac{g(U_0)}{g(U_0) - g\left(U^{*(1)}\right)} \qquad [7.15]$$

which is closer to the true failure surface than $U^{*(1)}$ ($g(U^{*(1)}) \neq 0$ most of the time).

All the points of the first NED are maintained in the second and $\left((M+1)(M+2)/2 - (M+1)\right)$ complementary points are added, in a half-star shape design around U_{C2}. Among the complementary points, M points are located according to [7.12] which implies that points are situated towards the failure region with respect to the central point and at a distance from the latter proportional to the local sensitivity of the RS. The $\left((M+1)(M+2)/2 - 2(M+1)\right)$ remaining complementary points are again generated from the M previous ones, each of them playing the role of a new local central point. When applying equation [7.12], the axes are considered in descending order with respect to the components of the gradient vector $\nabla \hat{g}(U_{C2})$. The point $U^{*(1)}$ is kept in the second NED under the condition:

$$\left\|U_{C2} - U^{*(1)}\right\| \leq h_0 \qquad [7.16]$$

The weighting factors are now computed by:

$$W_k = exp\left(-\frac{g\left(\boldsymbol{u}^{(k)}\right) - y_{min}}{g(\boldsymbol{U}_0)}\right) exp\left(-\frac{\left\|\boldsymbol{U}^{*(iter-1)} - \boldsymbol{u}^{(k)}\right\|^2}{2}\right) \qquad [7.17]$$

in which the last term allows the closeness to the previous most probable failure point to be accounted for. If the condition [7.16] is not fulfilled, $\boldsymbol{U}^{*(1)} = \boldsymbol{U}_{C2}$ in [7.17]. Once the unknown coefficients of $\hat{g}(\boldsymbol{U})$ are computed by equation [7.13], the second most probable failure point $\boldsymbol{U}^{*(2)}$ is determined by FORM.

7.5.5.1.3. Further iterations (iter>2)

The NED is enriched with the point $\boldsymbol{U}^{*(iter-1)}$. The weighting factors and the coefficients of $\hat{g}(\boldsymbol{U})$ are updated by equations [7.17] and [7.13] respectively. A new most failure point is determined. The convergence of the procedure is achieved when:

$$\left\{\begin{array}{l} \left|\beta^{(iter)} - \beta^{(iter-1)}\right| \leq \varepsilon_\beta \\ \left\|\boldsymbol{U}^{*(iter)} - \boldsymbol{U}^{*(iter-1)}\right\| \leq \varepsilon_P \end{array}\right. \qquad [7.18]$$

7.5.5.2. Examples

Two examples are reported here in order to show the interest but also the limit of the polynomial response surface method. The method described above is referred to as RSDW (Response Surface with Double Weighting).

7.5.5.2.1. Example 1

A simple explicit limit state function is considered in a standardized space:

$$g(\boldsymbol{U}) = exp(0.4(U_2 + 2) + 6.2) - exp(0.3U_1 + 5) - 200 \qquad [7.19]$$

The results are reported in Table 7.1 where N_r denotes the number of runs of the true limit state function. It can be seen that the values of the reliability index are very close to each other for all the methods under consideration. In the same way, the values of $\left|g(\boldsymbol{U}^*)/g(\boldsymbol{U}_0)\right|$ are close to zero and express a good closeness of the most probable point with regard to the failure surface. The cumulative formation of the RS is depicted in Figure 7.4. The effect on the quality and efficiency of the initial grid size h_0 can be seen in Tables 7.2 and 7.3. It is worth noting that the use of a weighting system allows this effect to be mitigated and a satisfactory closeness to the true failure surface to be obtained

Method	β	U^*_1	U^*_2	$\left\|g\left(U^*\right)/g\left(U_0\right)\right\|$	N_r
Adaptive MC[KAY 04]	2.710	0.969	-2.531	3.36×10^{-5}	-
RS [KIM 97]	2.691	-	-	-	-
RS [KAY 04]	2.686	0.820	-2.558	5.84×10^{-3}	8
RS [DUP 06]	2.710	0.951	-2.538	9.10×10^{-4}	21
RSDW ($R^2 = 0.997$)	2.707	0.860	-2.567	8.48×10^{-4}	12

Table 7.1. *Comparison between several RS methods (Example 1)*

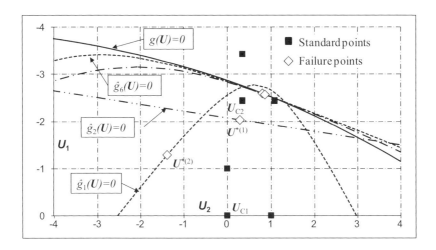

Figure 7.4. *Cumulative formation of the response surface (Example 1)*

h_0	1	2	3	4
β	2.707	2.724	2.713	2.714
$\left\|g\left(U^*\right)/g\left(U_0\right)\right\|$	8.48×10^{-4}	4.06×10^{-4}	8.02×10^{-4}	6.88×10^{-4}
R^2	0.997	0.999	0.999	0.930
N_r	12	10	11	13

Table 7.2. *Influence of h_0 with a weighting system ($W_k\neq1$) (Example 1)*

h_0	1	2	3	4
β	2.707	2.715	2.715	2.726
$\left| g(U^*)/ g(U_0) \right|$	1.15×10^{-3}	1.45×10^{-3}	1.12×10^{-3}	3.16×10^{-3}
R^2	0.997	0.999	0.999	0.885
N_r	14	11	11	15

Table 7.3. *Influence of h_0 without weighting system ($W_k=1$) (Example 1)*

u_{max} (cm)		4	5	6
β	RSDW	2.19	2.86	3.28
	FORM	2.20	2.86	3.40
$\left| g(U^*)/ g(U_0) \right|$	RSDW	5.39×10^{-3}	2.80×10^{-3}	1.06×10^{-3}
	FORM	6.03×10^{-6}	6.25×10^{-4}	1.72×10^{-5}
N_r	RSDW	256	258	255
	FORM	80	75	104
R^2		0.993	0.994	0.993

Table 7.4. *Comparison between RS and direct FORM (Example 2)*

7.5.5.2.2. Example 2

An implicit limit state function is considered which involves the top displacement of a multi-storey and multi-span steel frame. This example is also presented in Chapter 8 (section 8.4.2) where the distributions of 21 correlated variables can be found. The results are reported in Table 7.4, where u_{max} denotes the threshold value of the displacement. As far as the convergence and closeness to the failure surface are concerned, it can be noted that RSDW is satisfactory. From a comparison with direct FORM (direct coupling between the Rackwitz–Fiessler algorithm and the true limit state function) it can be said that, on one hand the RS method supplies consistent values of the reliability index but, on the other hand, the computational cost is significantly higher. For such an example, RSDW is therefore less efficient than direct FORM.

7.6. Conclusion

The response surface method was first developed for any system whose response to stimuli could not be satisfactorily captured by explanatory models, and was employed for biological systems. When applying the response surface method to mechanical systems, the main goal is to reduce the computational cost resulting, for example, from the use of finite element explanatory models. Some specific developments of the response surface method have been carried out in the context of probabilistic reliability analysis. The examples reported in this chapter show that these developments are not fully efficient in terms of computational cost if the number of variables exceeds twenty. Nonetheless, the quality of the adaptive response surface presented above is sufficiently good for the probabilistic results to be very close to those obtained by the use of the true limit state function. Moreover, having an explicit approximate failure function in the region where the most failure point is located facilitates probabilistic post-processing, including assessment of the failure probability, sensitivity analysis, etc.

7.7. Bibliography

[ANS 63] ANSCOMBE F.J., TUKEY F.W., "The examination and analysis of residuals", *Technometrics*, vol. 5, p.141-160, 1963.

[ATK 68] ATKINSON A.C., HUNTER W.G., "The design of experiments for parameter estimation", *Technometrics*, vol. 10, p.271-289, 1968.

[BIE 83] BIER V.M., "A measure of uncertainty importance for components in fault trees", *ANS Trans.*, 45(1), p.384-385, 1983.

[BLI 35a] BLISS C.I., "The calculation of the dosage-mortality curve", *Ann. Appl. Biol.*, vol. 22, p.134-167, 1935.

[BLI 35b] BLISS C.I., "The comparison of the dosage-mortality data", *Ann. Appl. Biol.*, vol. 22, p.307-333, 1935.

[BOU 95] BOUYSSY V., RACKWITZ R., "Approximation of non-normal responses for drag dominated offshore structures", *Proc. 6th IFIP WG 7.5*, "Reliability and Optimization of Structural Systems", no.7, Chapman and Hall, London, p.161-168, 1995.

[BOX 50] BOX G.E.P., WILSON K.B., "On the experimental attainment of optimum conditions", *J. R. Statist. Soc.*, B13, p.1-45, 1950.

[BOX 55] BOX G.E.P., YOULE P.V., "The exploration and exploitation of response surfaces: an example of the link between the fitted surface and the basic mechanism of the system", *Biometrics*, vol. 11, p.287-322, 1955.

[BOX 57] BOX G.E.P., HUNTER J.S., "Multifactor experimental designs for exploring response surfaces", *Ann. Math. Statist.*, vol. 28, p.195-241, 1957.

[BOX 59a] Box G.E.P., DRAPER N.R., "A basis for the selection of a response surface design", *J. Amer. Statist. Ass.*, vol. 54, p.622-654, 1959.

[BOX 59b] Box G.E.P., LUCAS H.L., "Design of experiments in nonlinear situations", *Biometrika*, vol. 46, p.77-90, 1959.

[BUC 90] BUCHER C.G., BOURGUND U., "A fast and efficient response surface approach for structural reliability problems", *Structural Safety*, vol. 7(1), p.57-66, 1990.

[CRO 41] CROWTHER E.M., YATES F., "Fertiliser policy in war-time", *Empire J. Exp. Agric.*, vol. 9, p.77-97, 1941.

[DEV 97] DEVICTOR N., MARQUES M., LEMAIRE M., "Utilisation des surfaces de réponse dans le calcul de la fiabilité des composants mécaniques", *Revue Française de Mécanique*, 1997-1, p.43-51, 1997.

[DIT 96] DITLEVSEN O., MADSEN H., *Structural Reliability Methods*, John Wiley & Sons, Chichester, 1996.

[DUP 06] DUPRAT F., SELLIER A., "Probabilistic approach to corrosion risk due to carbonation *via* an adaptive response surface method", *Probabilistic Engineering Mechanics*, 21(4), p.207-216, 2006.

[ELS 62] ELSTON R.C., GRIZZLE J.E., "Estimation of time response curves and their confidence bands", *Biometrics*, vol. 18, p.148-159, 1962.

[ENE 94] ENEVOLDSEN I., FABER M.H., SORENSEN J.D., "Adaptive response surface technique in reliability estimation", in SCHUËLLER G.I., SHINOZUKA M., & YAO J.T.P. (Eds), *Structural Safety and Reliability, Proceedings of the ICOSSAR'93*, Balkema Rotterdam Brookfield, p.1257-1264, 1994.

[FAV 89] FAVARELLI L., "Response surface approach for reliability analysis", *Journal of Engineering Mechanics*, vol. 115(12), p.2763-2781, 1989.

[GAD 33] GADDUM J.H., Reports on biological Standards III: methods on biologocal assay depending on a quantal response, Special Report No. 183, Medical Research Council, HMSO, London, 1933.

[GAV 08] GAVIN H.P., YAU S. C., "High-order limit state functions in the response surface method for structural reliability analysis", *Structural Safety*, vol. 30(2), p.162-179, 2008.

[GAY 03] GAYTON N., BOURINET J.M., LEMAIRE M., "CQ2RS: a new statistical approach to the response surface method for reliability analysis", *Struct. Safety*, vol. 25(1), p.99-121, 2003.

[GUA 01] GUAN X.L., MELCHERS R.E., "Effect of response surface parameter variation on structural reliability estimates", *Structural Safety*, vol. 23(4), p.429-444, 2001.

[GUP 04] GUPTA S., MANOHAR C.S., "An improved response surface method for the determination of failure probability and importance measures", *Structural Safety*, vol. 26(2), p.123-139, 2001.

[HEL 93] HELTON J.C., "Uncertainty and sensitivity analysis techniques for use in performance assessment for radioactive waste disposal", *Rel. Engrg. and Syst. Saf.*, vol. 42, p.327-367, 1993.

[IMA 87] IMAN R.L., "A matrix-based approach to uncertainty and sensitivity analysis for fault trees", *Risk Analysis*, vol. 7(1), p.21-33, 1987.

[JCS 02] JOINT COMMITTEE ON STRUCTURAL SAFETY, JCSS probabilistic model code, 2002, http ://www.jcss.ethz.ch.

[KAY 04] KAYMAZ I., MCMAHON C., "A response surface method based on weighted regression for structural reliability analysis", *Probabilistic Engineering Mechanics*, vol. 20(1), p.1-7, 2004.

[KHA 89] KHATIB-RAHBAR M., "A probabilistic approach to quantifying uncertainties in the progression of severe accidents", *Nucl. Sci. Engrg.*, vol. 10(2), p. 219-259, 1989.

[KIE 52] KIEFER J., WOLFOWITZ J., "Stochastic estimation of the maximum of a regression function", *Ann. Math. Statist.*, vol. 23, p.462-466, 1952.

[KIE 60] KIEFER J., WOLFOWITZ J., "The equivalence of two extremum problems", *Canadian J. Math.*, 12, p.363-366, 1960.

[KIM 97] KIM S.H., NA S.W., "Response surface method using vector projeted sampling points", *Structural Safety*, 19(1), p.3-19, 1997.

[LAB 95] LABEYRIE J., SCHOEFS F., "A discussion on response surface approximations for use in structural reliability", *6th IFIP WG7.5, Reliability and Optimisation of Structural Systems*, no. 15, Chapman & Hall, p.161-168, 1995.

[LAB 96] LABEYRIE J., SCHOEFS F., "Matrix response surfaces for describing environmental loads", Vol II Safety and Reliability*, Proc. of 15th Int. Conf. on Offshore Mechanics and Arctic Engineering*, (O.M.A.E'96), Florence, Italy, p.119-126, ASME 1996.

[LEM 09] LEMAIRE M., in collaboration with CHATEAUNEUF A. and MITTEAU J.C., *Structural Reliability*, ISTE Ltd, London and John Wiley & Sons Inc., New York, 2009.

[MEA 75] MEAD R., PIKE D.J., "A review of response surface methodology from a biometric viewpoint", *Biometrics* 31, p.803-851, 1975.

[MIT 30] MITSCHERLICH E.A., *Die Bestimmung des Dungerbedurfnisses des Bodens*, Paul Paray, Berlin, Germany, 1930.

[MUZ 93] MUZEAU J.P., LEMAIRE M., EL-TAWIL K., "Méthodes fiabilistes des surfaces de réponse quadratiques et évaluation des règlements", *Construction Métallique*, vol. 3, p.41-52, 1993.

[NAK 95] NAKAYASU H., "Relation between parameters sensitivities and dimensional invariance on Stochastic materials design of fibrous composite laminates", *6th IFIP WG 7.5, Reliability and Optimisation of structural system*, no. 21, p.209-216, 1995.

[NEL 65] NELDER J.A., MEAD R., "A simplex method for function minimization", *Comp. J. 7*, p.308-313, 1965.

[NEL 66] NELDER J.A., "Inverse polynomials, a useful group of multi-factor response functions", *Biometrics* 22, p. 128-141, 1966.

[NGU 09] NGUYEN X.S., SELLIER A., DUPRAT F., PONS G., "Adaptive response surface method based on a double weighted regression technique", *Probabilistic Engineering Mechanics*, vol. 24(2), p. 135-143, 2009.

[PAR 94] PARK C.K., ANH K.II., "A new approach for measuring uncertainty importance and distributional sensitivity in probabilistic safety assessment", *Rel. Engrg. and Syst. Saf.*, vol. 46, p. 253-261, 1994.

[POW 65] POWEL M.J.D., "A method for minimizing a sum of squares of nonlinear functions without calculating derivatives", *Comp. J.*, vol. 8, p. 303-307, 1965.

[RAJ 93] RAJASHEKHAR M.R., ELLINGWOOD B.R., "A new look at the response surface approach for reliability analysis", *Structural Safety*, vol. 12, p. 205-220, 1993.

[RAO 58] RAO C.R., "Some statistical methods for the comparison of growth curves", *Biometrics*, vol. 14, p. 1-17, 1958.

[REE 29] REED L.J., BERKSON J., "The application of the logistic function for experimental data", *J. Phys. Chem.*, vol. 33, p. 760–799, 1929.

[ROB 51] ROBINS H., MONRO S., "A stochastic approximation method", *Ann. Math. Statist.*, vol. 22, p.400-407, 1951.

[SCH 08] SCHOEFS F., "Sensitivity approach for modelling the environmental loading of marine structures through a matrix response surface", *Reliability Engineering and System Safety*, 93(7), July 2008, p.1004-1017, 2008, doi: dx.doi.org/10.1016/j.ress.2007.05.006.

[SCH 07] SCHOEFS F., Méthodologie des surfaces de réponse physiques pour une analyse de risque intégrée des structures offshore fissurées, Mémoire d'Habilitation à Diriger les Recherches, ED MTGC, University of Nantes, France, 3 December 2007.

[SCH 96] SCHOEFS F., Surface de réponse des efforts de houle dans le calcul de fiabilité des Ouvrages, PhD thesis, ED 82-208, University of Nantes, France, 21 November 1996.

[STE 51] STEVENS W.L., "Asymptotic regression", *Biometrics*, vol. 7, p. 247-267, 1951.

[WIN 32] WINSOR C.P., "The Gompertz curve as a growth curve", *Proc. Natl. Acad. Sci.*, vol. 18, p. 1-8, 1932.

[WIS 39] WISHART J., "Statistical treatment of animal experiment", *J. R. Statist. Soc.*, B6, p. 1-22, 1939.

[WON 05] WONG S.M., HOBBS R.E., ONOF C., "An adaptive response surface method for reliability analysis of structures with multiple loading sequences", *Structural Safety*, vol. 217(3), p. 287-308, 2005.

Chapter 8

Response Surfaces based on Polynomial Chaos Expansions

8.1. Introduction

8.1.1. *Statement of the reliability problem*

Let us recall the problem stated in the introduction to Chapter 1. Of interest is a building or a part of a building reduced to a mechanical structure, whose behavior may be represented by a mechanical model. The latter is described by a transfer function $\mathcal{M}$ (often known implicitly, e.g. under the form of a finite element code) that allows the effects of the loading (e.g. displacements, strains, stresses) to be evaluated, depending on input parameters which describe the structure and its environment, that is the geometrical properties (e.g. dimensions, cross-section areas and moments of inertia of the beam elements), the material properties (e.g. Young's modulus, Poisson's coefficient) and the loading (e.g. applied loads, thermal loading). All these input parameters are gathered in a vector x. The model response is denoted by $y = \mathcal{M}(x)$.

A *probabilistic model* is then defined for the input parameters. In this context, the latter are described by a *random* vector X (of size M) with a prescribed joint probability density function (PDF) of $f_X(x)$.

Lastly, a *limit state function* is defined which mathematically represents the failure criterion with respect to which the structure has to be assessed. This function

---

Chapter written by Bruno SUDRET, Géraud BLATMAN and Marc BERVEILLER.

which is denoted by $g(\mathcal{M}(X), X')$ accounts for the effects of the loading (e.g. displacements, stresses) to which (probabilistic or deterministic) limits – gathered in a vector X' – are assigned. It is conventionally defined in such a way that its negative values correspond to realizations x of the input parameters which lead to failure.

Denoting by $f_{X,X'}$ the joint PDF of X and X', the reliability analysis aims at evaluating the probability of failure P_f of the structure under consideration that reads:

$$P_f = \int_{\{x:g(\mathcal{M}(x),x')\leq 0\}} f_{X,X'}(x, x')\, dx \qquad\qquad [8.1]$$

Many methods may be used to solve this problem, such as Monte Carlo simulation, FORM/SORM methods, directional simulation, subset simulation; see for example the references Ditlevsen & Madsen [DIT 96], and Lemaire [LEM 09].

8.1.2. *From Monte Carlo simulation to polynomial chaos expansions*

The Monte Carlo method is well known in structural reliability and more generally in probabilistic mechanics. It relies upon the generation of a random sample of the input variables, denoted by $\mathcal{X} = \{x_i, i = 1, ..., N\}$. For each sample x_i, first the mechanical response $y_i = \mathcal{M}(x_i)$ then the limit state function is evaluated. The number of samples N_f that leads to a negative value of g is computed. Then the probability of failure is estimated by $\hat{p}_f = N_f / N$. The Monte Carlo method is easy to implement and also robust since it provides confidence intervals for the estimate $\hat{p}_f$. However it is computationally very expensive, especially when low probabilities of failure are sought (with orders of magnitude ranging from 10^{-3} to 10^{-6} in practice). Indeed, it is shown that an accurate estimation (say, with a relative accuracy of 5%) of a probability of magnitude 10^{-k} requires about $N = 4.10^{k+2}$ points in the sample set.

From another point of view, the Monte Carlo method consists of characterizing the random response of a structure $Y = \mathcal{M}(X)$ *pointwise* in its domain of variation, i.e. from a given set of random realizations of Y. Thus a large number of simulations is expected in order to accurately estimate the probabilistic content of Y, e.g. through its PDF $f_Y(y)$ which may be estimated by the histogram of the sample $\mathcal{Y} = \{\mathcal{M}(x_i), i = 1, ..., N\}$. In an industrial context, most models are of the finite element type and necessitate a significant CPU time (say from a few minutes to a few hours), hence this approach cannot be applied.

As an alternative, Y can be considered intrinsically as a random variable belonging to a specific space (such as the space of random variables with a finite variance), and can be represented in a suitable basis for this space. Thus the response is cast as a converging series, as follows:

$$Y \approx \sum_{j=0}^{+\infty} a_j \Psi_j \qquad [8.2]$$

where $\{\Psi_j, j \in \mathbb{N}\}$ is a set of random variables that form the basis and where $\{a_j, j \in \mathbb{N}\}$ is the set of the "coordinates" of Y in this basis. In particular, a special focus is given to bases made of orthonormal polynomials of random variables. The series in [8.2] is then referred to as *polynomial chaos* (PC) *expansion*.

In the remainder of this chapter, the building of a PC basis (section 8.2), then the computation of the PC coefficients and their post-processing, dedicated to reliability analysis (section 8.3) are each described. Lastly, two application examples are addressed in section 8.4.

8.2. Building of a polynomial chaos basis

8.2.1. *Orthogonal polynomials*

For the sake of simplicity, the input random variables are assumed to be independent. Their marginal PDF is denoted by $f_i(x_i)$, thus their joint PDF reads $f_X(x) = \prod_{i=1}^{M} f_i(x_i)$. For each input random variable X_i, a family of orthonormal polynomials $\{P_j^i, j \in \mathbb{N}\}$ can be defined, such that $P_0^i \equiv 1$ and the degree of each polynomial P_j^i is j, $j > 0$. The orthonormality property is defined by:

$$< P_j^i, P_k^i >\equiv \int_{D_i} P_j^i(x) P_k^i(x) f_i(x) \, dx = \delta_{j,k} \qquad [8.3]$$

where $\delta_{j,k} = 1$ if $j = k$ and 0 otherwise, and D_i is the support of the random variable X_i. In practice, classical families of orthonormal polynomials can be associated with usual continuous random variables. If X_i is Gaussian, the corresponding family is that made of Hermite polynomials. If it uniformly distributed, the corresponding family is that made of Legendre polynominals [ABR 70], [SCH 00].

Then a basis made of *multivariate* polynominals $\{\Psi_j, j \in \mathbb{N}\}$ can be easily built up by *tensorization*, that is, by multiplying the univariate polynomials as follows:

$$\psi_\alpha(x) \equiv \psi_{\alpha_1,...,\alpha_M}(x) = P^1_{\alpha_1}(x_1) \times \cdots \times P^M_{\alpha_M}(x_M) \qquad [8.4]$$

It has been shown by Soize & Ghanem [SOI 04] that the family $\{\Psi_\alpha \equiv \psi_\alpha(X), \alpha \in \mathbb{N}^M\}$ form an appropriate countable basis to represent the random response $Y = \mathcal{M}(X)$ of a mechanical model. In addition, this basis is *orthonormal* with respect to the inner product in the space of random variables, with a finite variance defined by the mathematical expectation $< Y_1, Y_2 >= \mathbb{E}[Y_1 Y_2]$. Indeed, from equations [8.3] and [8.4], we get:

$$< \Psi_\alpha, \Psi_\beta >= \mathbb{E}[\psi_\alpha(X)\psi_\beta(X)] = \delta_{\alpha\beta} \qquad [8.5]$$

The elements of the basis (indexed by their multi-index α) are classically ordered according to their increasing total degree $p = |\alpha| = \sum_{i=1}^{M} \alpha_i$, and are enumerated from $j = 0$ to infinity, as in equation [8.2] (an algorithm allowing a systematic building of the basis may be found in [SUD 06]).

In practice, it is necessary to retain only a finite number of terms in the PC basis. Then the series is generally truncated in such a way that only those basis polynomials ψ_j with a total degree not greater than a given p are retained. Hence, the truncated series containing P terms is:

$$Y \approx \mathcal{M}^{PC}(X) \equiv \sum_{j=0}^{P-1} a_j \psi_j(X) \qquad [8.6]$$

where it is shown that $P = \binom{M+p}{p}$.

8.2.2. *Example*

Let us consider the random response Y of a mechanical model $Y = \mathcal{M}(X_1, X_2)$ depending on two Gaussian random variables X_1 and X_2, with mean value μ_i and standard deviation σ_i, $i = 1,2$. When applying the linear mapping $X_i = \mu_i + \sigma_i \xi_i$, the response can be recast in terms of standard Gaussian random variables, that is $Y = \tilde{\mathcal{M}}(\xi_1, \xi_2)$.

The family of orthogonal polynomials with respect to the standard Gaussian PDF $\varphi(x) = 1/\sqrt{2\pi}e^{-x^2/2}$ is the family of Hermite polynomials $\{He_j(x), j \in \mathbb{N}\}$. They are defined by the following recurrence relationship:

$$
\begin{aligned}
He_{-1}(x) \equiv He_0(x) &= 1 \\
He_{n+1}(x) &= x\,He_n(x) - n\,He_{n-1}(x)
\end{aligned}
$$

[8.7]

The resulting polynomials are orthogonal but not orthonormal. They have various specific properties, as shown in [BAR 05]. In particular, it is shown that $<He_n, He_n> = n!$. Therefore the family $\{He_j(x)/\sqrt{n!}, j \in \mathbb{N}\}$ is orthonormal. The four first normalized Hermite polynomials are thus $\{1, x, (x^2 - 1)/\sqrt{2}, (x^3 - 3x)/\sqrt{6}\}$.

Assume that the expansion of the random response Y onto a PC basis of maximal degree $p = 3$ is of interest. The retained polynomials are built from products of Hermite polynomials in ξ_1 and ξ_2 (Table 8.1). Hence an approximation of the model response (a stochastic response surface) is sought under the form:

$$
\begin{aligned}
Y \equiv \tilde{\mathcal{M}}^{PC}(\xi_1, \xi_2) &= a_0 + a_1 \xi_1 + a_2 \xi_2 + a_3 (\xi_1^2 - 1)/\sqrt{2} + a_4 \xi_1 \xi_2 \\
&+ a_5 (\xi_2^2 - 1)/\sqrt{2} + a_6 (\xi_1^3 - 3\xi_1)/\sqrt{6} + a_7 (\xi_1^2 - 1)\xi_2/\sqrt{2} \\
&+ a_8 (\xi_2^2 - 1)\xi_1/\sqrt{2} + a_9 (\xi_2^3 - 3\xi_2)/\sqrt{6}
\end{aligned}
$$

[8.8]

where the coefficients $\{a_j, j = 0, ..., 9\}$ must be determined.

8.3. Computation of the expansion coefficients

8.3.1. *Introduction*

Polynomial chaos expansions were originally introduced to represent random fields [WIE 38]. They have been used more recently for solving Stochastic Partial Differential Equations (SPDE) [GHA 91]. In this setup, investigations have been conducted in many fields such as biology, mechanics, fluid mechanics and thermal physics [GHA 98], [ISU 98], [KNI 06], [SUD 04], [WIN 85], [XIU 03]. The weak formulation of these SPDEs is discretized both in the physical space (e.g. by finite elements) and in the probabilistic space (e.g. onto the PC basis). The coefficients arising from equation [8.6] are obtained by a Galerkin method [GHA 91], and are obtained by solving a large system of coupled linear equations, which may reveal time and memory consumption [PEL 00]. This method is referred to as *intrusive* due to the coupled nature of the system. The application of intrusive spectral methods to structural reliability analysis were initially proposed in [SUD 00], [SUD 02].

j	α	$\Psi_\alpha \equiv \Psi_j$
0	[0,0]	$\Psi_0 = 1$
1	[1,0]	$\Psi_1 = \xi_1$
2	[0,1]	$\Psi_2 = \xi_2$
3	[2,0]	$\Psi_3 = \left(\xi_1^2 - 1\right)/\sqrt{2}$
4	[1,1]	$\Psi_4 = \xi_1\xi_2$
5	[0,2]	$\Psi_5 = \left(\xi_2^2 - 1\right)/\sqrt{2}$
6	[3,0]	$\Psi_6 = \left(\xi_1^3 - 3\xi_1\right)/\sqrt{6}$
7	[2,1]	$\Psi_7 = \left(\xi_1^2 - 1\right)\xi_2/\sqrt{2}$
8	[1,2]	$\Psi_8 = \left(\xi_2^2 - 1\right)\xi_1/\sqrt{2}$
9	[0,3]	$\Psi_9 = \left(\xi_2^3 - 3\xi_2\right)/\sqrt{6}$

Table 8.1. *Examples of polynomial chaos of degree 3 with 2 input variables* (ξ_1,ξ_2)

On the other hand, *non intrusive* methods have recently received increasing interest. They allow the coefficients in equation [8.6] to be computed by means of a set of *deterministic* calculations, i.e. a set $\mathcal{Y} = \{\mathcal{M}(x_i), i = 1,..., N\}$ of evaluations of the model response at suitably chosen values of the input variables. The *non-intrusive* label indicates that these methods can be applied using the deterministic code associated with the model $\mathcal{M}$ without modification.

Two classes of approach may be distinguished among the non intrusive methods, namely the *projection* approach [BAR 05], [BAR 06], [BLA 07], [LEM 01] and the *regression* approach [BER 05], [BER 06]. These are detailed below in turn.

8.3.2. Projection methods

The so-called *projection methods* take benefit of the orthonormality of the PC basis. Indeed, by multiplying the expansion [8.2] by $\psi_j(X)$ and by integrating with respect to the joint PDF $f_X(x)$ of X, we achieve:

$$a_j = \mathbb{E}\left[\mathcal{M}(X)\psi_j(X)\right] \equiv \int_{D_X} \mathcal{M}(x)\psi_j(x)f_X(x)\,dx \qquad [8.9]$$

In practice, the above expression is estimated using classical methods for numerical integration, which consist of approximating the multi-dimensional integral by a weighted sum, as follows:

$$\hat{a}_j \approx \sum_{i=1}^{N} w_i\,\mathcal{M}(x_i)\psi_j(x_i) \qquad [8.10]$$

Several techniques can be considered which differ from the choice of the integration points x_i and weights w_i.

The so-called *simulation* method relies upon the choice of N *random* integration points and integration weights equal to $1/N$, which leads to:

$$\hat{a}_j \approx \frac{1}{N}\sum_{i=1}^{N} \mathcal{M}(x_i)\psi_j(x_i) \qquad [8.11]$$

This corresponds to the application of Monte Carlo simulation to the estimation of the expectation in [8.9]. The accuracy of the coefficient estimators depends on the sampling strategy adopted. In case of a standard random sample (classical Monte Carlo simulation), a relatively low convergence rate in $N^{-1/2}$ is obtained. The convergence speed may be increased using stratified sampling techniques, such as latin hypercube sampling [MCK 79]. Moreover, it is shown that the use of quasi-random numbers [NIE 92], which are generated from deterministic low discrepancy sequences, guarantees a better filling of the domain of variation of the parameters and lead to faster convergences [BLA 07].

As an alternative, the integral in equation [8.9] can be approximated by a Gauss quadrature scheme. Its principle is well known in the unidimensional case: the integral of a function $h(x)$ (weighted by a function $w(x)$) is estimated by a sum of evaluations of h in a set of quadrature points:

$$I = \int_D h(x)w(x)\,dx \approx \sum_{k=1}^{n} w_i\,h(x_i) \qquad [8.12]$$

The Gauss method allows us to integrate exactly any polynomial function of a degree not greater than $2n-1$ with n suitable integration points, namely the roots of the orthogonal polynomials with respect to the weight function $w(x)$ in the sense of equation [8.3]. The extension to the multi-dimensional case (integral [8.9]) is obtained by tensorizing the univariate quadrature rules:

$$a_j \approx \sum_{i_1=1}^{n_1} \cdots \sum_{i_M=1}^{n_M} w_{i_1}^1 \cdots w_{i_M}^M \, \psi_j \left(x_{1,i_1}, ..., x_{M,i_1} \right) \mathcal{M} \left(x_{1,i_1}, ..., x_{M,i_1} \right) \qquad [8.13]$$

Isotropic formulae are commonly used, that is formulae which satisfy $n_1 = ... = n_M = n$. It is shown that this scheme allows us to integrate exactly any multivariate polynomial of a partial degree not greater than $2n-1$. Now, if the model response is approximated by a PC expansion of degree p, then the integrand in [8.9] is a polynomial of total degree $2p$. Therefore, a tensorized quadrature rule with $n = p+1$ points is used in order to estimate the coefficients. Such a strategy leads to the performance of $N = (p+1)^M$ model evaluations, which may prove to be cumbersome in the presence of a large number of input parameters (say, $M \geq 5-8$).

The computational effort may be dramatically reduced by replacing the full tensor product [8.13] with the so-called Smolyak scheme [SMO 63], also known as sparse quadrature. This technique has been applied in relation to PC expansions in [KEE 03], [SUD 07].

8.3.3. *Regression methods*

8.3.3.1. *Direct approach*

An alternative method to projection consists of computing the coefficients which provide the best approximation of $Y = \mathcal{M}(X)$ in the least squares sense by a truncated PC expansion containing a fixed number P of terms. Using the following vector notation:

$$a = \{a_0, ..., a_{P-1}\}^{\mathrm{T}} \qquad [8.14]$$

$$\psi(X) = \{\psi_0(X), ..., \psi_{P-1}(X)\}^{\mathrm{T}} \qquad [8.15]$$

equation [8.6] is re-written as:

$$\mathcal{M}^{PC} = a^{\mathrm{T}} \, \psi(X) \qquad [8.16]$$

Let us consider a set $\mathcal{X} \equiv \{x_1, ..., x_N\}^{\mathrm{T}}$ of realizations of the input random vector, which is an experimental design. Let us denote the set of corresponding model evaluations by $\mathcal{Y}$. The experimental design may be built either from a random or a quasi-random sample (see section 8.3.2), or from the roots of the orthogonal polynomials that are used to build the basis [BER 05].

The problem consists of finding the vector of coefficients $\hat{a}$ that minimize the sum of squared errors (see Chapter 1, sections 1.6–1.7); that is:

$$\hat{a} = \arg\min_a \sum_{i=1}^{N} \left(a^{\mathrm{T}} \psi(x_i) - \mathcal{M}(x_i) \right)^2 \qquad [8.17]$$

It is shown that the solution can be obtained in closed form as follows:

$$\hat{a} = \left(\Psi^{\mathrm{T}} \Psi \right)^{-1} \Psi^{\mathrm{T}} \mathcal{Y} \qquad [8.18]$$

where the generic entry of matrix Ψ is given by:

$$\Psi_{ij} = \psi_j(x_i) \quad i = 1, ..., N \quad j = 0, ..., P-1 \qquad [8.19]$$

It is necessary that the number N of model evaluations be greater than the number P of unknown coefficients in order to make the problem "well-posed". In the case of a random or a latin hypercube [MCK 79] experimental design, the rule-of-thumb that $N = 2P$ generally leads to satisfactory results. As shown in [BER 05], [BER 06], regression methods appears to be particularly efficient at computing PC coefficients. It also allows us to define *a posteriori* error estimates as well as an adaptive strategy for building the PC basis, which is outlined in the following sections [BLA 09].

8.3.3.2. *Error estimation*

The approximation error of a PC expansion can be quantified by the coefficient of determination, R^2, which is currently used in regression analysis. This coefficient depends on the sum of squared deviations between the "true" model response and the PC representation:

$$R^2 \equiv 1 - \frac{1/N \sum_{i=1}^{N} \left(\mathcal{M}(x_i) - \mathcal{M}^{PC}(x_i) \right)^2}{\hat{V}[Y]} \qquad [8.20]$$

where $\hat{V}[Y]$ is the empirical variance of the model evaluations:

$$\hat{V}[Y] \equiv \frac{1}{N-1} \sum_{i=1}^{N} (y_i - \bar{y})^2 \qquad \bar{y} = \frac{1}{N} \sum_{i=1}^{N} y_i \tag{8.21}$$

Thus, $R^2 = 1$ corresponds to a perfect adequation, whereas $R^2 = 0$ indicates a very poor approximation. However, the R^2 coefficient should be used with caution, as it tends to under-predict the genuine approximation error. In the extreme case of $N = P$, the PC approximation interpolates the model realizations, which leads to $R^2 = 1$ even if the error may reveal significant points that do not belong to the experimental design $\mathcal{X}$. This phenomenon is known as *overfitting*.

As a consequence, a more robust error estimate is used, which is based on a cross validation technique named *leave-one-out* [ALL 71], [SAP 06]. In this setup, for any point in the experimental design x_i, we compute the deviation $\Delta^{(i)}$ between the observation $y_i = \mathcal{M}(x_i)$ and the evaluation in x_i of a PC expansion denoted by $\mathcal{M}_{\mathcal{X}\backslash x^{(i)}}^{PC}$, whose coefficients are computed from the experimental design $\mathcal{X}\backslash\{x_i\}$ obtained by removing the point x_i from $\mathcal{X}$. By analogy with the R^2 coefficient, the Q^2 coefficient are defined as follows:

$$Q^2 \equiv 1 - \frac{1/N \sum_{i=1}^{N} \Delta^{(i)\,2}}{\hat{V}[Y]} \qquad \Delta^{(i)} = \mathcal{M}(x_i) - \mathcal{M}_{\mathcal{X}\backslash x_i}^{PC}(x_i) \tag{8.22}$$

8.3.3.3. *Adaptive approach*

Section 8.3.3.1 has shown that the size N of the experimental design has to be greater than the number P of terms in the truncated PC series in order to solve the regression problem. Now, P strongly increases with both the maximal degree p of the PC expansion and the number M of input random variables, according to the formula $P = \binom{M+p}{p} = \frac{(M+p)!}{M!\,p!}$. Thus the regression method may lead to intractable calculations in high dimensions (say $M \geq 10$). In order to reduce the number of model evaluations, a sparse PC approximation of the response Y is sought, that is, a PC representation which only contains a small number of nonzero coefficients. Of course it is not possible to determine *a priori* the significant terms. Hence an iterative procedure has been proposed in [BLA 08], [BLA 09], [BLA 10b] to build up a sparse PC expansion step-by-step. The algorithm is outlined in Figure 8.1.

First, an initial experimental design $\mathcal{X}$ is considered and the associated model evaluations are gathered in $\mathcal{Y}$. The model response is approximated by a PC expansion of degree $p = 0$ (i.e. the current basis is a constant term). Then, terms corresponding to polynomials with increasing degree p and interaction order j are proposed. Two steps can be distinguished:

– a forward step: all the candidate terms are added in turn to the current basis. The changes in R^2 due to the addition of each term are evaluated. Eventually all those terms which lead to a significant increase of R^2 are retained;

– a backward step: all the terms in the current basis are removed in turn, and the associated changes in R^2 are computed. Eventually all those terms which lead to an insignificant decrease of R^2 are discarded.

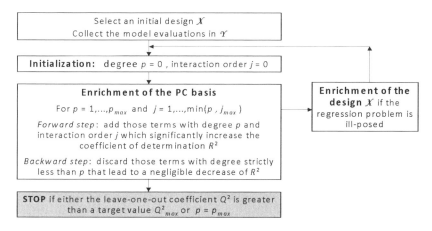

Figure 8.1. *Computational flowchart of the procedure to build up an adaptive sparse polynomial chaos expansion*

As well as this adaptation of the PC basis, the experimental design is systematically enriched in such a way that the various regression problems are always well-posed. For this purpose, sequential sampling strategies are adopted which are based either on quasi-random numbers or nested LHS (Latin Hypercube Sampling) [BLA 10b], [WAN 03]. The algorithm stops when the Q^2 coefficient related to the current PC approximation has reached a target accuracy Q^2_{tgt}.

8.3.4. *Post-processing of the coefficients*

As mentioned previously, the random variable $Y = \mathcal{M}(X)$ is thoroughly defined by its coefficients a_j which can be estimated by means of several non-intrusive methods. In particular, the mean and the variance Y can be derived analytically from these coefficients due to the orthonormality of the basis:

$$E[Y] = a_0 \qquad V[Y] = \sum_{j=1}^{P-1} a_j^2 \tag{8.23}$$

For reliability analyses, the model response is substituted by a PC decomposition into the limit state function which describes the system failure. Considering, for simplicity, a failure criterion associated with a deterministic maximum admissible threshold y_{max}, the limit state function reads:

$$g(X) = y_{\text{max}} - \mathcal{M}(X) \tag{8.24}$$

Substituting the model response $\mathcal{M}(X)$ by a PC expansion $\mathcal{M}^{PC}(X)$ into equation [8.24], we get the *analytical* limit state function:

$$g^{PC}(X) = y_{\text{max}} - \sum_{j=0}^{P-1} a_j \psi_j(X) \tag{8.25}$$

This quantity corresponds to a *stochastic response surface* which replaces the original limit state function, that is, an analytical (polynomial) expression whose evaluation cost is negligible. Thus the probability of failure may be estimated inexpensively by applying the classical reliability methods (e.g. direct Monte Carlo simulation, FORM and importance sampling).

It should be noted that the PC-based approximation [8.25] differs from the quadratic response surfaces used in reliability analysis, which are *local* approximations, i.e. in the vicinity of the design point when using FORM (see Chapter 1 for more details). In this context, a parametric study (e.g. changing the threshold y_{max} in equation [8.24]) leads to a new response surface being built for each value of the parameter, in contrast to the PC approach.

8.4. Applications in structural reliability

8.4.1. *Elastic engineering truss*

8.4.1.1. *Problem outline*

We will now consider an elastic engineering truss, represented in Figure 8.2 [BLA 07], [SUD 07]. Ten input random variables are used, whose distribution, means and standard deviations are reported in Table 8.2. The quantity which is of interest is the (random) maximum vertical displacement of the structure, denoted by $V_1 = \mathcal{M}(X)$.

8.4.1.2. *Reliability analysis*

Of interest is the reliability of the truss structure with respect to an admissible maximal displacement. The associated limit state function reads:

$$g(X) = v_{max} - |\mathcal{M}(X)| \leq 0, \, v_{max} = 11 \, cm \qquad\qquad [8.26]$$

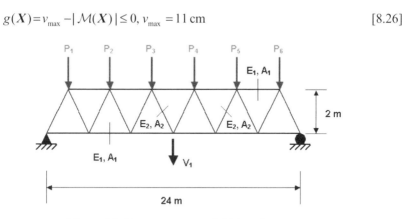

Figure 8.2. *Truss structure with 23 bar elements*

Variable	Distribution	Mean	Standard Deviation
E_1, E_2 (MPa)	LogNormal	210,000	21,000
A_1 (cm²)	LogNormal	20	2
A_2 (cm²)	LogNormal	10	1
P_1- P_6 (kN)	Gumbel	50	7.5

Table 8.2. *Elastic truss – random input variables*

The reference value of the probability of failure is obtained by direct Monte Carlo simulation using $N = 10^6$ samples. The reliability analysis is carried out from various PC approximations of the response (denoted by $\mathcal{M}^{PC}(X)$) made of normalized Hermite polynomials. To this end, the input random vector $X = \{E_1, E_2, A_1, A_2, P_1, ..., P_6\}^T$ is transformed into a random vector containing 10 independent standard Gaussian variables.

A PC expansion of degree 3 is considered. The PC coefficients are computed by Smolyak sparse quadrature (1,771 calculations were performed). The reference results are obtained by applying an importance sampling strategy (the importance PDF is centered on the design point determined by FORM), from which a generalized reliability index is obtained. The results are gathered in table 8.3. A 2% accuracy on the reliability index is obtained for all the admissible thresholds in the interval [10–16] cm.

The results obtained from a PC expansion whose coefficients have been estimated by regression are reported in Table 8.4. Precisely, the reliability indices are alternatively computed from a full PC expansion of degree $p = 3$ and from a sparse PC representation. Whatever the approach, the coefficients have been calculated from an experimental design made of quasi-random numbers. It appears that both the full and the sparse PC approximation yield accurate estimates of β, with a relative error less than 3.5%.

Threshold (cm)	Reference		Smolyak Projection	
	P_f	β	P_f	β
10	4.31 x 10^{-2}	1.715	4.29 x 10^{-2}	1.718
11	8.70 x 10^{-3}	2.378	8.73 x 10^{-3}	2.377
12	1.50 x 10^{-3}	2.967	1.47 x 10^{-3}	2.974
14	3.49 x 10^{-5}	3.977	2.83 x 10^{-5}	4.026
16	6.03 x 10^{-7}	4.855	4.01 x 10^{-7}	4.935

Table 8.3. *Elastic engineering truss – reliability results obtained using a third order polynomial chaos expansion based on a Smolyak Projection*

Threshold (cm)	Reference	Full PC		Sparse PC	
	β^{REF}	$\hat{\beta}$	ε (%)	$\hat{\beta}$	ε (%)
10	1.715	1.71	0.6	1.72	0.0
11	2.378	2.38	0.0	2.38	0.0
12	2.967	2.98	0.3	2.99	0.7
14	3.977	4.04	1.5	4.07	2.3
16	4.855	4.95	2.1	5.02	3.5
Error $1 - Q^2$		1 x 10^{-6}		9 x 10^{-5}	
Number of terms		286		114	
Number of model evaluations		443		207	

Table 8.4. *Elastic engineering truss – reliability results obtained using a full (with p=2) and a sparse PC expansion based on regression*

It is observed that the sparse PC approach only requires half as many calculations as the full PC approach based on regression. It only requires one eight as many calculations as the full PC approach based on Smolyak quadrature).

8.4.2. *Frame structure*

The frame structure represented in Figure 8.3 is now considered [BLA 10b], [LIU 91]. The frame beam elements are made of 8 different materials, whose properties are gathered in Table 8.5.

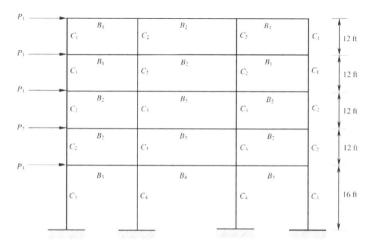

Figure 8.3. *Example of a frame structure subjected to lateral loads*

Element	Young's modulus	Moment of inertia	Cross-sectional area
B_1	E_4	I_{10}	A_{18}
B_2	E_4	I_{11}	A_{19}
B_3	E_4	I_{12}	A_{20}
B_4	E_4	I_{13}	A_{21}
C_1	E_5	I_6	A_{14}
C_2	E_5	I_7	A_{15}
C_3	E_5	I_8	A_{16}
C_4	E_5	I_9	A_{17}

Table 8.5. *Frame structure: element properties*

The response of interest is the horizontal component u of the top-floor displacement at the top right corner. The 3 applied loads, the 2 Young's moduli, the 8 moments of inertia and the 8 cross-section areas of the frame components are assumed to be random. They are collected in the random vector $X = \{P_1, P_2, P_3, ..., I_6, ..., I_{13}, A_{14}, ..., A_{21}\}^T$ of size $M = 21$. The properties of the random variables are reported in Table 8.6.

Variable	Distribution	Mean †	Standard Deviation †
P_1 (kN)		133.454	40.04
P_2 (kN)	Lognormal	88.970	35.59
P_3 (kN)		71.175	28.47
E_1 (kN/m²)	Truncated normal over	2.17375×10^7	1.9152×10^6
E_2 (kN/m²)	$[0, +\infty)$	2.37964×10^7	1.9152×10^6
I_6 (m⁴)		8.13443×10^{-3}	1.08344×10^{-3}
I_7 (m⁴)		2.13745×10^{-2}	2.59609×10^{-3}
I_8 (m⁴)		2.59610×10^{-2}	3.02878×10^{-3}
I_9 (m⁴)	Truncated normal over	1.08108×10^{-2}	2.59610×10^{-3}
I_{10} (m⁴)	$[0, +\infty)$	1.41055×10^{-2}	3.46146×10^{-3}
I_{11} (m⁴)		2.32785×10^{-2}	5.62487×10^{-3}
I_{12} (m⁴)		2.59610×10^{-2}	6.49024×10^{-3}
I_{13} (m⁴)		2.13745×10^{-2}	2.59609×10^{-3}
A_{14} (m²)		3.12564×10^{-1}	5.58150×10^{-2}
A_{15} (m²)		3.72100×10^{-1}	7.44200×10^{-2}
A_{16} (m²)		5.06060×10^{-1}	9.30250×10^{-2}
A_{17} (m²)	Truncated normal over	5.58150×10^{-1}	1.11630×10^{-1}
A_{18} (m²)	$[0, +\infty)$	2.53020×10^{-1}	9.30250×10^{-2}
A_{19} (m²)		2.91168×10^{-1}	1.02323×10^{-1}
A_{20} (m²)		3.73030×10^{-1}	1.20933×10^{-1}
A_{21} (m²)		4.18600×10^{-1}	1.95375×10^{-1}

† The mean value and the standard deviation of the cross-sections, moments of inertia and Young's moduli are those of the untruncated Gaussian distributions

Table 8.6. *Frame structure: random input variables*

Moreover, the random input variables are correlated as follows: the correlation coefficients between the cross-section areas and the moments of inertia of a given element are equal to $\rho_{A_i I_i} = 0.95$, the correlation coefficients related to the other geometrical properties are equal to $\rho_{A_i I_j} = \rho_{I_i I_j} = \rho_{A_i A_j} = 0.13$, the correlation coefficient between the two Young's moduli is equal to $\rho_{E_4 E_5} = 0.9$. The random vector X is transformed into a vector of independent standard Gaussian random variables by means of a Nataf transform $X = T(\xi)$ [NAT 62] prior to building the PC expansion.

Let us study the serviceability of the frame structure with respect to the limit state function:

$$g(X) = u_{\max} - \mathcal{M}(T(\xi)) \qquad [8.27]$$

where $u_{\max}$ is a maximal admissible horizontal displacement. It is approximated by an analytical function by replacing the model $\mathcal{M}(T(\xi))$ with its PC representation made of Hermite polynomials, denoted by $\mathcal{M}^{PC}(\xi)$. A parametric study is carried out varying the threshold u_{max} from 4 to 8 cm. The generalized reliability indices are estimated by post-processing a full third-order PC as well as a sparse PC. The results are reported in Table 8.7.

Threshold (cm)	Reference	Full PC		Sparse PC	
	β^{REF}	$\hat{\beta}$	ε (%)	$\hat{\beta}$	ε (%)
4	2.27	2.26	0.4	2.29	0.9
5	2.96	3.00	1.4	3.01	1.7
6	3.51	3.60	2.6	3.61	2.8
7	3.96	4.12	4.0	4.11	3.8
8	4.33	4.58	5.8	4.56	5.3
Error $1 - Q^2$		1×10^{-3}		1×10^{-3}	
Number of terms		2,024		138	
Number of model evaluations		3,724		450	

Table 8.7. *Frame structure: estimates of the generalized reliability index $\beta = -\Phi^{-1}(P_f)$ for various values of threshold displacement*

As observed in the truss example, the estimation error of the reliability index slightly increases with the threshold value. Both types of PC approximation yield relative errors for β of less than 5% when the threshold ranges from 4 to 8 cm. The sparse PC approach reveals much more efficiently than the full PC approach, with a gain factor of 8 in terms of the number of model evaluations (only 450 finite element runs instead of 3,724).

8.5. Conclusion

The methods based on polynomial chaos expansions have motivated many investigations over the last few years. Their application to structural reliability is quite novel and can be viewed as a particular type of stochastic response surface (for application in geotechnical engineering, see [SUD 08]). This chapter has shown the principles of these methods with respect to the simulation techniques classically used in reliability analysis, and has introduced their formalisms.

The two application examples have shown the interest in using PC-based response surfaces for reliability analysis. Indeed, with a computational cost varying from 200 model evaluations for the elastic truss involving 10 input random parameters) to 450 (for the frame structure involving 21 input random parameters), the probabilities of failure associated to various thresholds were obtained. In practice, the parametric study is carried out at a negligible cost with respect to a single reliability analysis, as the polynomial chaos expansion is built once and for all, and is then post-processed for the various threshold values.

In addition to probabilities of failure, polynomial chaos expansion can also be used in order to study the sensitivity of a response, by evaluating its probability density function, its statistical moments and the sensitivity indices to the input variables, still without requiring any additional model evaluation [BAR 07], [SUD 07]. [BAR 07] has also shown that the use of a basis made of Lagrange polynomials may be a relevant alternative to Hermite polynomials when the number of input parameters remains low (say $M < 4$-5).

Finally, the building of adaptive sparse polynomial bases paves the way to the solving of high-dimensional problems ($M \sim 50$–100) at a reasonable computational cost ($N < 1,000$), in particular thanks to the introduction in probabilistic mechanics of advanced statistical regression methods such as Least Angle Regression (LAR) [BLA 11].

8.6. Bibliography

[ABR 70] ABRAMOWITZ M., STEGUN I., *Handbook of Mathematical Functions*, Dover Publications, Inc., Mineola, NY, USA, 1970.

[ALL 71] ALLEN D., The prediction sum of squares as a criterion for selecting prediction variables, Report no. 23, Dept. of Statistics, University of Kentucky, USA, 1971.

[BAR 05] BAROTH J. Analyse par EFS de la propagation d'incertitudes à travers un modèle mécanique non linéaire, Thesis, University of Clermont II, France, 2005.

[BAR 06] BAROTH J., BODÉ L., BRESSOLETTE P., FOGLI M., "SFE method using Hermite Polynomials: an approach for solving nonlinear problems with uncertain parameters", *Comp. Meth. Appl. Mech. Engrg.*, 195, p. 6479-6501, 2006.

[BAR 07] BAROTH J., BRESSOLETTE P., CHAUVIERE C., FOGLI M., "An efficient SFE method using Lagrange Polynomials: application to nonlinear mechanical problems with uncertain parameters", *Comp. Meth. Appl. Mech. Engrg.*, 196, p. 4419-4429, 2007.

[BER 05] BERVEILLER M., Eléments finis stochastiques: approches intrusive et non intrusive pour des analyses de fiabilité, PhD thesis, Blaise Pascal University, Clermont-Ferrand, France, 2005.

[BER 06] BERVEILLER M., SUDRET B., LEMAIRE M., "Stochastic finite elements: a non intrusive approach by regression", *Eur. J. Comput. Mech.*, vol. 15(1-3), p. 81-92, 2006.

[BLA 07] BLATMAN G., SUDRET B., BERVEILLER M., "Quasi-random numbers in stochastic finite element analysis", *Mécanique & Industries*, vol. 8, p. 289-297, 2007.

[BLA 08] BLATMAN G., SUDRET B., "Sparse polynomial chaos expansions and adaptive stochastic finite elements using a regression approach", *Comptes Rendus Mécanique*, vol. 336(6), p. 518-523, 2008.

[BLA 09] BLATMAN G., Adaptive sparse polynomial chaos expansions for uncertainty propagation and sensitivity analysis, PhD thesis, Blaise Pascal University, Clermont-Ferrand, France, 2009.

[BLA 10a] BLATMAN G., SUDRET B., "Efficient computation of Sobol' sensitivity indices using sparse polynomial chaos expansions", *Reliab. Eng. Sys. Safety*, vol. 95(11), p. 1216-1229, 2010.

[BLA 10b] BLATMAN G., SUDRET B., "An adaptive algorithm to build up sparse polynomial chaos expansions for stochastic finite element analysis", *Prob. Eng. Mech.*, vol. 25(2), p. 183-197, 2010.

[BLA 11] BLATMAN G., SUDRET B., "Adaptive sparse polynomial chaos expansions based on Least Angle Regression", *J. Comput. Phys.*, vol. 230(6), p. 2345-2367, 2011.

[DIT 96] DITLEVSEN O., MADSEN H., *Structural Reliability Methods*, John Wiley & Sons, Chichester, 1996.

[GHA 91] GHANEM R., SPANOS P., *Stochastic Finite Elements – A Spectral Approach*, Springer Verlag, New York, USA, 1991 (Re-issued by Dover Publications, 2003).

[GHA 98] GHANEM R., DHAM D., "Stochastic finite element analysis for multiphase flow in heterogeneous porous media", *Transp. Porous Media*, vol. 32, p. 239-262, 1998.

[ISU 98] ISUKAPALLI S.S., ROY A., GEORGOPOULOS P.G., "Stochastic response surface methods for uncertainty propagation: application to environmental and biological systems", *Risk analysis*, 18(3), p. 351-362, 1998.

[KEE 03] KEESE A., MATTHIES H., "Sparse quadrature as an alternative to Monte Carlo for stochastic finite element techniques", *Proc. Appl. Math. Mech.*, vol. 3(1), p. 493-494, 2003.

[KNI 06] KNIO O., LE MAÎTRE O., "Uncertainty propagation in CFD using polynomial chaos decomposition", *Fluid Dyn. Res.*, vol. 38(9), p. 616-640, 2006.

[LEM 01] LE MAÎTRE O., KNIO O., NAJM H., GHANEM R., "A stochastic projection method for fluid flow – I. Basic formulation", *J. Comput. Phys.*, vol. 173, p. 481-511, 2001.

[LEM 09] LEMAIRE M., *Structural Reliability*, ISTE Ltd, London and John Wiley & Sons, New York, USA, 2009.

[LIU 91] LIU P.-L., DER KIUREGHIAN A., "Optimization algorithms for structural reliability", *Structural Safety*, vol. 9, p. 161-177, 1991.

[MCK 79] MCKAY M. D., BECKMAN R. J., CONOVER W. J., "A comparison of three methods for selecting values of input variables in the analysis of output from a computer code", *Technometrics*, vol. 2, p. 239–245, 1979.

[NAT 62] NATAF A., "Détermination des distributions dont les marges sont données", *C. R. Acad. Sci. Paris*, vol. 225, p. 42-43, 1962.

[NIE 92] NIEDERREITER H., *Random Number Generation and Quasi-Monte Carlo Methods*, Society for Industrial and Applied Mathematics, Philadelphia, PA, USA, 1992.

[PEL 00] PELLISSETTI M.-F., GHANEM R., "Iterative solution of systems of linear equations arising in the context of stochastic finite elements", *Adv. Eng. Soft.*, vol. 31, p. 607-616, 2000.

[SAP 06] SAPORTA G., *Probabilités, analyse des données et statistique* (2nd edition), Editions Technip, Paris, France, 2006.

[SCH 00] SCHOUTENS W., *Stochastic Processes and Orthogonal Polynomials*, Springer-Verlag, New York, USA, 2000.

[SMO 63] SMOLYAK S., "Quadrature and interpolation formulas for tensor products of certain classes of functions", *Soviet. Math. Dokl.*, vol. 4, p. 240-243, 1963.

[SOI 04] SOIZE C., GHANEM R., "Physical systems with random uncertainties: chaos representations with arbitrary probability measure", *SIAM J. Sci. Comput.*, vol. 26(2), p. 395-410, 2004.

[SUD 00] SUDRET B., DER KIUREGHIAN A., Stochastic finite elements and reliability: a state-of-the-art report, Report issue UCB/SEMM-2000/08, 173 p., University of California, Berkeley, USA, 2000.

[SUD 02] SUDRET B., DER KIUREGHIAN A., "Comparison of finite element reliability methods", *Prob. Eng. Mech.*, vol. 17, p. 337-348, 2002.

[SUD 04] SUDRET B., BERVEILLER M., LEMAIRE M., "A stochastic finite element method in linear mechanics", *Comptes Rendus Mécanique*, vol. 332, p. 531-537, 2004.

[SUD 06] SUDRET B., BERVEILLER M., LEMAIRE M., "A stochastic finite element procedure for moment and reliability analysis", *Eur. J. Comp. Mech.*, vol. 15(7-8), p. 825-866, 2006.

[SUD 07] SUDRET B., Uncertainty propagation and sensitivity analysis in mechanical models – Contributions to structural reliability and stochastic spectral methods, Habilitation à diriger des recherches, Blaise Pascal University, Clermont-Ferrand, France, 2007.

[SUD 08] SUDRET B., BERVEILLER M., "Stochastic finite element methods in geotechnical engineering", in *Reliability-based design in geotechnical engineering: computations and applications*, K.K. PHOON (Ed.), 2008.

[WAN 03] WANG G., "Adaptive response surface method using inherited Latin Hypercube design points", *J. Mech. Design*, vol. 125, p. 210-220, 2003.

[WIE 38] WIENER N., "The homogeneous chaos", *Amer. J. Math.*, vol. 60, p. 897-936, 1938.

[WIN 85] WINTERSTEIN S.R., "Non normal responses and fatigue dommage", *J. of Engrg. Mech.*, vol. 111(10):1291-1295, 1985.

[XIU 03] XIU D., KARNIADAKIS G., "Modelling uncertainty in steady state diffusion problems *via* generalized polynomial chaos", *Comput. Methods Appl. Mech. Engrg.*, vol. 191(43), p. 4927–4948, 2003.

PART 4

Methods for Structural Reliability over Time

Introduction to Part 4

In order to predict the behavior and the degree of reliability of a component or a structure, we require a precise understanding of the properties of the materials and the actions that are applied to them. Reliability must be ensured throughout the entire service life of the structure. Moreover, civil engineering typically involves very long service lives for its structures (several tens or even hundreds of years). As a result, the potential evolution of properties over time must also be appreciated, whether this is through "natural" ageing, the response of materials to extreme environmental loads, or simply fatigue caused by normal actions.

Thus, this part of the book will consider the changes in state of various structures over time, by presenting and applying a range of methods.

Chapter 9 presents and uses data aggregation and unification to evaluate the evacuation time of a building following the outbreak of a fire. Chapter 10 introduces probabilistic methods used for reliability studies of mechanical problems over time, illustrated using the example of the engineering truss already considered in Part 3 but in this case subject to a time-varying load. Chapter 11 considers Monte Carlo methods using Markov chains and includes an example which considers the lifetime of a system in series, in the case of sparse and censored data. Finally, Chapter 12 considers reliability updating using real-world feedback, presenting an example which considers the prediction of creepage in the pressure vessel of a nuclear power station.

Chapter 9

Data Aggregation and Unification

9.1. Introduction

The data collected to evaluate failure scenarios for civil engineering works may take a variety of different forms: experimental data, expert opinion, output from mechanical models, data from auscultation. They may be deterministic or random. It is generally difficult to simultaneously manipulate data in disparate formats. This course of action is, however, essential when there is insufficient information of a given type available to carry out a detailed study, or when there is a need to obtain a "rapid" estimate of the consequences of failure scenarios from the available information; this same approach applies to a brief criticality analysis.

9.2. Methods of data aggregation and unification

Data aggregation and unification methods manipulate heterogeneous data, giving consideration to their format, their level of precision and their relevance to the questions to be answered in terms of quantification of scenario consequences. Because of this, it is crucial to evaluate the relative quality of these data.

This information on quality is referred to as the belief mass and it can be evaluated using a data quality analysis scheme such as NUSAP (Numerical Unit Spread Assessment and Pedigree). This is a means of representing their scientific information content, taking into account the various qualitative and quantitative aspects of their uncertainty. It was introduced in 1990 by Funtowicz & Ravertz [FUN 90] and is still used today [CUR 08], [LAI 00], [SLU 07], [TAL 06a].

Chapter written by Daniel Boissier and Aurélie Talon.

This tool is built around the key concept of qualifying the five criteria contained within the NUSAP acronym: *Numerical, Unit, Spread, Assessment* and *Pedigree*. The first two criteria, "Numerical" and "Unit" refer to the quantitative aspect of the information. This could for example be a service life of 60, along with its units: years. The three other criteria represent the qualitative aspects of uncertainty [SLU 96]:

– "spread" represents the uncertainty (± 10 years); this uncertainty could be represented in terms of fuzzy subsets, probability distributions, etc.;

– "assessment" expresses a judgment on the reliability, representing the strength of the data point: it symbolizes the state-of-the-art of production science;

– "pedigree" represents the quality of the evaluation process used on the information, and indicates the "scientific status" of the available knowledge, in other words the context in which it is used, and the general culture from which the data point has been drawn.

Data unification and aggregation methods exploit the complementary nature of multi-source data. The choice of method – unification or aggregation of the data – will depend on the problem to be solved and on the available data:

– application of unification: when several different data sets are available from various sources, all of which relate to the same problem;

– application of aggregation: when several different data sets are available relating to different sub-problems of the problem to be solved.

We can also apply first one and then the other method if several data sets are available from different sources and refer to different sub-problems of the problem to be solved.

Consider, for example, the following problem: "Determine the evacuation time from an apartment by a wheelchair-bound individual following an outbreak of fire". The problem can be broken down into four sub-problems:

1: "Determine the time to perceive the alarm";

2: "Determine the time to move to the elevator";

3: "Determine the time to descend in the elevator";

4: "Determine the time to exit the elevator and leave the building".

Let us assume that we have ten pieces of data available (Table 9.1). Figure 9.1 shows a diagram representing how we can tackle the problem using these data.

Sub-problem	Data		
Perceive alarm	D_1. Statistics	D_2. Data from constructor	D_3. Comments from apartment occupant
Move to elevator	D_4. Theoretical model	D_5. Comments from apartment occupant	
Descend in elevator	D_6. Statistics	D_7. Data from constructor	D_8. Comments from elevator maintenance staff
Exit elevator and leave building	D_9. Theoretical model	D_{10}. Comments from apartment occupant	

Table 9.1. *Time data for the evacuation problem*

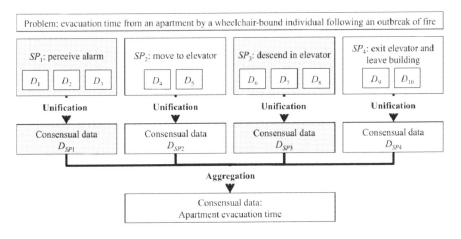

Figure 9.1. *Approach used to solve evacuation problem*

9.2.1. *Data unification methods*

Data unification refers to data fusion, drawing on Dempster–Shafer theory [SHA 76].

We will consider two pieces of data (or *evidence*) D_1 and D_2 belonging to two subsets S_1 and S_2 of the frame of discernment, with which are associated the belief masses $m(S_1)$ and $m(S_2)$, determined from the data quality analysis. Then, the fusion of these two data aims to determine firstly the belief masses that can be attributed to the intersection of these two subsets and to what is not known, and secondly the value of their consensus.

Table 9.2 shows the principle of mass allocation (Dempster–Shafer theory) for concordant data D_1 and D_2. We will not discuss the case of discordant evidence, which is discussed in [SHA 76].

	S_1	$S_1 \cap S_2 \neq \varnothing$	S_2	Ignorance	
Data D_1	$m(S_1)$	$m(S_1)$	$1 - m(S_1)$	$1 - m(S_1)$	
Data D_2	$1 - m(S_2)$	$m(S_2)$	$m(S_2)$	$1 - m(S_2)$	
Consensus	$m(S_1) \times (1 - m(S_2))$	$m(S_1) \times m(S_2)$	$(1 - m(S_1)) \times m(S_2)$	$(1 - m(S_1)) \times (1 - m(S_2))$	$\Sigma = 1$

Table 9.2. *Mass allocation for two concordant pieces of evidence*

The consensus of a subset (S_1, S_2, $S_1 \cap S_2$ and Θ/ignorance) is equal to the product of the belief masses assigned to this subset by each of the pieces of evidence.

The fusion (or combination) of n pieces of evidence is obtained by performing n-1 successive fusions of two pieces of evidence.

This can be formalized as follows:

$$(((D_1 \oplus D_2) \oplus D_3)...) \oplus D_n \qquad [9.1]$$

where $\oplus$ represents the operation of combining two pieces of evidence.

As noted by J. Lair [LAI 00], this approach (Dempster's rule of combination) is unsatisfactory if two strongly conflicting pieces of evidence are being combined. This leads to inconsistencies, which is particularly unfortunate since data collected in construction contexts can involve strong conflicts.

In order to remedy this limitation, J. Lair studied various alternative data assembly rules and proposed assembly strategies, a selection algorithms to determine the assembly rules that are most appropriate to the data being combined (extract shown in Figure 9.2).

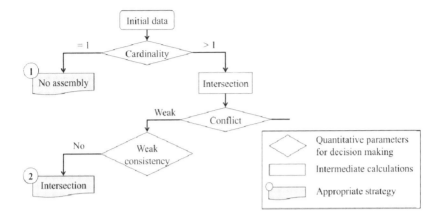

Figure 9.2. *Assembly strategy selection algorithm (from [LAI 00])*

The cardinality is the number of pieces of data available to be combined. Data are in conflict when they have a null intersection. The conflict between two pieces of data (subsets) S_1 and S_2 is defined by [SHA 76] through the equation:

$$Conf(S_1, S_2) = \ln\left(\frac{1}{1-m_c}\right)$$

[9.2]

where m_c represents the mass assigned to the empty set.

Fusion example: suppose we have three data to combine, D_1, D_2 and D_3, all of which are durations. The frame of discernment associated with these three pieces of data is $\Theta = [0; 60]$ min.

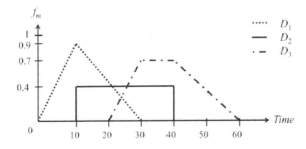

Figure 9.3. *Example of three pieces of evidence to be combined*

Preparation of these pieces of data leads to the results shown in Figure 9.4. The data preparation involves transforming a possibility distribution into a set of "nested" intervals using α-cuts. The aim of this preparation is to avoid centering the entire belief mass on a single point or on too narrow an interval, which could result in a conflict during the data fusion phase. For example, it could be difficult to achieve a consensus when the output from the preparation procedure leads to a significant and similar belief mass for two intervals of [5; 7] min and [15; 16] min, since their intersection is the empty set.

D_1	
Interval	Mass
[5; 20]	0.45
[0; 30]	0.45
Θ = [0; 60]	0.10
Total:	1

D_2	
Interval	Mass
[10; 40]	0.40
Θ = [0; 60]	0.60
Total:	1

D_3	
Interval	Mass
[25; 50]	0.35
[20; 60]	0.35
Θ = [0; 60]	0.30
Total:	1

Figure 9.4. *Results of pieces of data preparation for the three data in Figure 9.3*

We now combine data D_1 and D_2 in accordance with Dempster's rule (Figure 9.5).

	Interval	Mass				Interval	Mass
			[10; 20] ([5; 20]∩[10; 40])	[5; 20]		[5; 20]	0.27
	[5; 20]	0.45	**0,18** (0.45×0.40)	0.27		[10; 20]	0.18
D_1			[10; 30]	[0; 30]		[10; 30]	0.18
	[0; 30]	0.45	0.18	0.27		[0; 30]	0.27
			[10; 40]	[0; 60]		[10; 40]	0.04
	[0; 60]	0.10	0.04	0.06		[0; 60]	0.06
	Mass	0.40	0.60			Total:	1
	Interval	[10; 40]	[0; 60]				
		D_2					

Figure 9.5. *Fusion of data D_1 and D_2 by intersection (see Table 9.2)*

We then combine the result of this first fusion with data D_3, according to Dempster's intersection rule (Figure 9.6).

$D_1 \oplus D_2$ Interval	Mass			
[5; 20]	0.27	{Empty}	{20}	[5; 20]
		0.0945	0.0945	0,.81
[10; 20]	0.18	{ Empty }	{20}	[10; 20]
		0.063	0.063	0.054
[10; 30]	0.18	[25; 30]	[20; 30]	[10; 30]
		0.063	0.063	0.054
[0; 30]	0.27	[25; 30]	[20; 30]	[0; 30]
		0.0945	0.0945	0.081
[10; 40]	0.04	[25; 40]	[20; 40]	[10; 40]
		0.014	0.014	0.012
[0; 60]	0.06	[25; 50]	[20; 60]	[0; 60]
		0.021	0.021	0.018
	Mass	0.35	0.35	0.30
	Interval	[25; 50]	[20; 60]	[0; 60]
		D_3		

Interval	Mass
{20}	0.1575
[0; 30]	0.0810
[5; 20]	0.0810
[10; 20]	0.0540
[10; 30]	0.0540
[10; 40]	0.0120
[20; 30]	0.1575
[20; 40]	0.0140
[20; 60]	0.0210
[25; 30]	0.1575
[25; 40]	0.0140
[25; 50]	0.0210
[0; 60]	0.0180
{Empty}	**0.1575**
Total:	1

Figure 9.6. *Fusion of the result of $D_1 \oplus D_2$ with data D_3 by intersection (Table 9.2)*

The conflict resulting from this fusion is equal to:

$$Conf(D_1, D_2, D_3) = \ln\left(\frac{1}{1 - 0.1575}\right) = 0.17 \qquad [9.3]$$

The conflict is in this case fairly weak and the consistency between the three data is not weak, and so in accordance with the algorithm in Figure 9.2 we follow strategy number 2, indicating that the fusion of these three data is complete. The result of this fusion is the set of intervals and their associated belief weights, which are listed in Figure 9.6.

9.2.2. Data aggregation methods

Data aggregation involves making best use of the set of available data for individual sub-problems of the problem to be solved. Aggregation enables us to obtain consensual data and an indication of the quality of the result. Data aggregation varies depending on the type of data to be aggregated, whether these be durations, probabilities or values of a particular criterion. Rather than providing an exhaustive list of all existing aggregation methods without further discussion, we will discuss just one such method here in detail: duration aggregation.

Duration aggregation involves obtaining the duration of a problem (scenario), and its accompanying quality, using the durations of each sub-problem (or phenomenon) that makes up the problem, the qualities associated with these durations and the transition values between each sub-problem [TAL 06b], [TAL 07]. This principle of duration aggregation is shown in Figure 9.7 for our problem ("determine the evacuation time from an apartment by a wheelchair-bound individual following an outbreak of fire") which was outlined at the start of section 9.2.

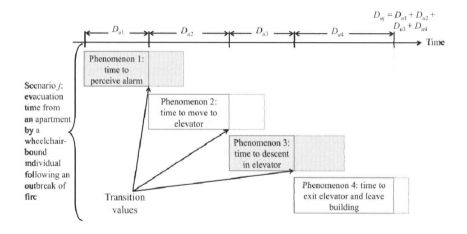

Figure 9.7. *Illustration of the principle of data aggregation*

The duration of each phenomenon can be given in the form of a discrete value or in the form of an interval. The quality associated with the duration of each phenomenon is:

– its belief mass, obtained by evaluating the quality of this data, in the case where a single data point is available to characterize this phenomenon;

– its Smets probability (hypothesis proposed in [TAL 06a]) as defined in section 9.3 (below), in the case where several data are available to characterize this phenomenon.

Let:

– Du^j, Du_i be the duration for scenario j (to be calculated) and for phenomenon i;

– n be the number of phenomena making up scenario j;

– m^j, m_i be the quality for scenario j (to be calculated) and the known qualities for phenomenon i, equal either to its belief mass (one data point available) or to its Smets probability (more than one data point available).

Then:

$$Du^j = \sum_{i=1}^{n} Du_i \qquad\qquad [9.4]$$

$$m^j = \min_{i \in \{1,\ldots,n\}} (m_i) \qquad\qquad [9.5]$$

Equation [9.5] is inspired by the addition formula for fuzzy subsets given in [BOU 95].

9.3. Evaluation of evacuation time for an apartment in case of fire

There are four main quantities that can be used to characterize the results of data unification and aggregation: consensus, plausibility, belief and Smets probability. The consensus sums up the belief masses associated with the set of intervals resulting from the fusion. casein the example we considered earlier, the intervals obtained from the fusion operations, along with their belief masses, are listed in the table on the right of Figure 9.6. The consensus curve is a simple and rapid means of determining a consensual duration and indications of conflict and ignorance in a data set: it is defined by:

$$C_f(\theta) = \sum_{\theta \in R_l} m_c(R_l) \qquad\qquad [9.6]$$

where θ is a fuzzy subset and R_l is the set of intervals resulting from data fusion. The consensus curve for the result of combining the three data in Figure 9.3 is shown in Figure 9.8.

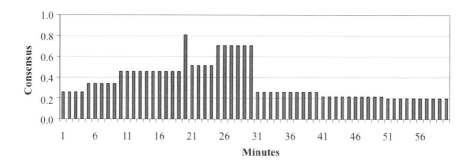

Figure 9.8. *Consensus curve for the three data in Figure 9.3*

If we consider Figure 9.8, we can conclude that there are two intervals that represent the maximum areas of consensus, the singleton of 20 years and the interval [25; 30] years. The existence of this singleton ({20} years) is biased by the fact that the Dempster intersection method concentrates the mass, "logically" assigned to ignorance, in singletons.

The belief associated with a subset θ, written $Bel(\theta)$, sums up all the reasons to support θ. It is the sum of the belief masses m_c associated with the subsets strictly included in θ; we can formalize this as follows:

$$Bel(\theta) = \sum_{R_l \subseteq \theta} m_c(R_l) \qquad [9.7]$$

The cumulative sum of the belief masses resulting from the fusion of the three pieces of data in Figure 9.3 is shown in Figure 9.9. The plausibility of a subset θ, written $Pl(\theta)$, expresses the strength with which we should support the subset θ. It corresponds to the sum of the belief masses associated with the non-null intersections of the subsets R_l and θ, in other words:

$$Pl(\theta) = \sum_{\theta \cap R_l \neq \phi} m_c(R_l) \qquad [9.8]$$

The cumulative sum of the plausibilities resulting from the fusion of the three data in Figure 9.3 is shown in Figure 9.9. The (belief, plausibility) pair is approximated by a probability that divides equally the mass assigned to each R_l, other than the singleton, among the R_l that comprise it [INR 03]. This is the Smets probability [DUB 90], or pignistic probability, written $P_S(\theta)$ when it is associated with a subset θ; this is defined by:

$$P_S(\theta) = \sum_{\theta \in R_l} \frac{m_c(R_l)}{|R_l|}$$ [9.9]

The belief, the Smets probability and the plausibility of a subset θ satisfy: $Bel(\theta) \leq P_S(\theta) \leq Pl(\theta)$.

As can be seen from Figure 9.9, the cumulative sums of the plausibility, the Smets probability and the belief are not exactly 1: this is due to Dempster's intersection rule, which assigns a mass to the empty set in case of conflict between data.

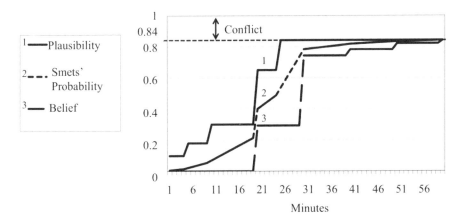

Figure 9.9. *Cumulative sums of belief, Smets probability and plausibility resulting from fusion of the three data shown in Figure 9.3*

Starting with these four quantities (consensus, plausibility, Smets probability and belief) that characterize the data fusion results, there are several different approaches for interpreting the results. We will discuss only one such approach here: obtaining an interval of values (for example, an interval of durations) for a given value of the plausibility, Smets probability and belief. Lair [LAI 00] uses these three quantities (plausibility, Smets probability and belief) to determine a characteristic service life with a probability of $k\%$, along with its uncertainty interval (plausible characteristic service life; credible characteristic service life), by plotting the cumulative distributions of these three quantities (like those shown in Figure 9.9 for the three data in Figure 9.3). This percentage indicates that $k\%$ of observations should have a duration less than the characteristic service life, with a probability of $k\%$.

For example, we estimate that failure may be observed (Figure 9.9):

– in less than 19 min with 40% plausibility;

– in less than 20 min with 40% Smets probability;

– in less than 29 min with 40% belief.

Further to the evaluation of the quality of the fusion result using plausibility indicators, Smets probability and belief, Lair [LAI 00] also proposed a method that can be used to evaluate the quality of the fusion procedure itself, with the help of a grid (Figure 9.10).

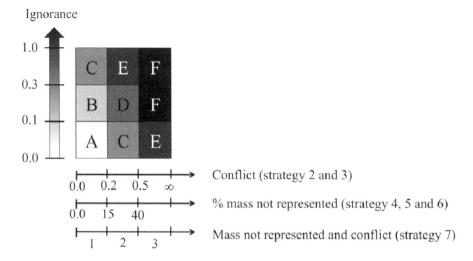

Figure 9.10. *Quality grid for the fusion process [LAI 00]*

The grid in Figure 9.10 represents the quality of the fusion process in the form of a letter, with A being the best and F the worst, as a function of the ignorance resulting from this fusion and the conflict it engenders (if strategy 2 or 3 was used), the percentage of mass not represented (if strategy 4, 5 or 6 was used), or the unrepresented mass and the conflict (if strategy 7 was used). These numbered strategies can be found in the full selection algorithm from which Figure 9.2 is extracted. For example, in the case of fusion of the three data in Figure 9.2, strategy 2 is followed, the ignorance is 0.018 (weight assigned to the set [0; 60] min – see Figure 9.6) and the conflict is 0.17. As a result, this fusion is represented by the quality letter A.

9.4. Conclusion

Data unification and aggregation methods enable the manipulation of a set of heterogeneous data (in terms of their format, origin and scale), and can also be used to evaluate the quality of the underlying data in order to obtain an estimate for the confidence that can be ascribed to the final result. Data unification can be used to manage a set of data from different sources in order to obtain consensual data from them. Data aggregation enables the scale of the analysis to be changed, for example obtaining an evaluation of the service life of a structure from the service lives of its components. It is possible to use the consensus, the plausibility, the Smets probability and the belief to interpret the result output from a data unification or aggregation process.

9.5. Bibliography

[BOU 95] BOUCHON-MEUNIER B., *La logique floue et ses applications*, Addison-Wesley, Paris, France, 1995.

[CUR 08] CURT C., Evaluation de la performance des barrages en service basée sur une formalisation et une agrégation des connaissances, PhD thesis, University of Clermont-Ferrand II, France, 2008.

[DUB 90] DUBOIS D., PRADE H., "Consonant approximations of belief functions", *International Journal of Approximate Reasoning*, vol. 4 p. 419-449, 1990.

[FUN 90] FUNTOWICZ S.O., RAVETZ J.R., *Uncertainty and Quality in Science for Policy*, Kluwer Academic Publishers, Dordrecht, The Netherlands, 1990.

[INR 03] INSTITUT NATIONAL DE LA RECHERCHE AGRONOMIQUE, Gestion de l'incertitude pour le diagnostic par le modèle de croyance transférée [online] 2003, see: http://www2.laas.fr/asinbio/SteyerLardon.pdf (accessed 27.04.2009).

[LAI 00] LAIR J., Evaluation de la durabilité des systèmes constructifs du bâtiment, civil engineering thesis, CSTB & LERMES, Clermont-Ferrand, France, 2000.

[SHA 76] SHAFER G., *A Mathematical Theory of Evidence*, Princeton University Press, Chichester, 1976.

[SLU 07] SLUIJS VAN DER J.P., "Uncertainty and precaution in environmental management: Insights from the UPEM conference", *Environmental Modelling & Software*, vol. 22, p. 590-598, 2007.

[SLU 96] SLUIJS VAN DER J.P., "Integrated assessment models and the management of uncertainties", *International Institute for Applied Systems Analysis*, Laxenburg, Austria, 1996.

[TAL 07] TALON A., BOISSIER D., LAIR J., "Service life assessment of building components: Application of evidence theory", *Canadian Journal of Civil Engineering*, vol. 35(3), p. 66-70, 2007.

[TAL 06a] TALON A., Evaluation des scénarios de dégradation des produits de construction, Civil engineering thesis, Centre scientifique et technique du bâtiment – division environnement, produits et ouvrages durables et Laboratoire génie civil, Clermont-Ferrand, France, 2006.

[TAL 06b] TALON A., BOISSIER D., HANS J., CAPRA B., "A multi-scale approach for service life evaluation", *European Symposium on Service Life and Serviceability of Concrete Structures*, Helsinki, Finland, 2006.

Chapter 10

Time-Variant Reliability Problems

10.1. Introduction

Although more or less ignored so far in this book, the time dimension is often present in structural reliability problems and has to be properly taken into account. Let us go back to the most basic formulation known as "*R-S*", in which failure occurs when a demand *S* is greater than a capacity *R*. It is clear here that for real structures both quantities may depend on time. Indeed:

– the resistance (or capacity) *R* of the structure (i.e. its material properties) may be degrading in time. Degradation mechanisms usually present an initiation phase and a propagation phase. Examples of such mechanisms are crack initiation and propagation in fracture mechanics, corrosion of steel structures and reinforced concrete rebars, decrease in steel toughness under irradiation in nuclear components, concrete creep and shrinkage, etc.;

– the load effect (or demand) *S* may randomly vary in time due to the time variation of the loading, e.g. environmental loads (wind velocity, temperature, wave height, etc.) or service loads (traffic, occupancy loads, etc.).

Both types of time dependency may be present simultaneously or not, and their nature is different: while degradation phenomena are usually monotonic and irreversible (corresponding to a decrease of resistance), loads are usually "oscillating" in nature and should be modeled by random processes.

Chapter written by Bruno SUDRET.

The aim of this chapter is not to fully cover the theory and tools of time-variant reliability problems, which is beyond the scope of the book. In contrast, it aims to define the basic concepts and focuses on a specific approach known as "PHI2 method", which allows the analyst to solve time-variant reliability problems using time-invariant tools such as the FORM method. For a more complete treatment of time-variant reliability problems, readers are referred to the numerous publications by Rackwitz [RAC 01], [RAC 04] and the books by Ditlevsen & Madsen ([DIT 96], Chapter 15) and Melchers ([MEL 9], Chapter 6).

10.2. Random processes

Random processes allow us to mathematically describe loads that are randomly varying in time [CRA 67], [LIN 67]. In the sections below, the basic notions are introduced without too much mathematical rigor.

10.2.1. *Definition and elementary properties*

A random process $X_t(\omega)$ is a set of random variables indexed by the time instant $t \in [0,T]$ with values in $\mathcal{D}_X \subset \mathbb{R}$. In this notation $\omega \in \Omega$ denotes the elementary events of an abstract probability space $(\Omega, \mathcal{F}, \mathbb{P})$. At each time instant the process reduces to a random variable $X_{t_0}(\omega)$ which is assigned some prescribed distribution. Conversely, a *realization* or *trajectory* of the process corresponds to the usual function $t \to X_t(\omega_0)$ for a given ω_0. This is simply denoted by small letters, say $x(t, \omega_0)$. In order to define a random process completely, the full set of joint probability distribution functions of any finite subset of random variables $\{X_{t_1}(\omega), \cdots, X_{t_N}(\omega)\}$ for any time instants $0 \leq t_1 < \cdots < t_N \leq T$ are prescribed. For structural reliability purposes, however, specific types of processes are of common use, e.g. Poisson, rectangular renewal wave or Gaussian processes, whose description is much easier, as seen below.

The usual definitions of marginal probability density functions (PDFs), statistical moments (mean value $\mu_X(t)$, standard deviation $\sigma_X(t)$, etc.) that are already well known for random variables naturally exist for random processes "at each time instant". Also of crucial importance is the autocorrelation function, defined as follows:

$$R_{XX}(t_1, t_2) = \mathrm{E}\left[X_{t_1} X_{t_2} \right]$$

[10.1]

where $E[.]$ denotes the mathematical expectation. This function represents the statistical dependence of points of trajectories considered at time instants t_1, t_2. Similarly, the autocorrelation coefficient function is defined by:

$$\rho_{XX}(t_1, t_2) = \frac{E\left[X_{t_1} X_{t_2}\right] - \mu_X(t_1) \mu_X(t_2)}{\sigma_X(t_1) \sigma_X(t_2)} \tag{10.2}$$

Loosely speaking, a random process is said to be stationary if its "characteristics" are invariant in time. Various rigorous definitions may be given. We will limit ourselves here to second-order stationarity, which implies that the statistical moments $E\left[X_t^k\right], k = 1, 2$ do not depend on time and that the autocorrelation function is invariant under time shift: $R_{XX}(t_1 + h, t_2 + h) = R_{XX}(t_1, t_2)$. The latter equation implies that the autocorrelation function only depends on the time interval $\tau = t_2 - t_1$.

A random process is said to be differentiable if the following limit, $\frac{X_{t+h}(\omega) - X_t(\omega)}{h}$, exists in the mean-square sense. The limit process is denoted by $\dot{X}_t$ and satisfies:

$$\lim_{h \to 0^+} E\left[\left(\frac{X_{t+h} - X_t}{h} - \dot{X}_t\right)^2\right] = 0 \tag{10.3}$$

Due to linearity, the mean value of the derivative process is equal to $\mu_{\dot{X}}(t) = \frac{d\mu_X(t)}{dt}$. It may easily be shown that its autocorrelation function reads:

$$R_{\dot{X}\dot{X}}(t_1, t_2) = \frac{\partial^2 R_{XX}(t_1, t_2)}{\partial t_1 \partial t_2} \tag{10.4}$$

In particular, for a stationary process, the following relationship holds:

$$R_{\dot{X}\dot{X}}(\tau) = -\frac{d^2 R_{XX}(\tau)}{d\tau^2}$$

10.2.2. *Gaussian random processes*

In contrast to other fields (e.g. in quantitative finance), the random processes that are used in engineering in order to model time-varying loads (wind velocity, wave height, etc.) show some regularity that is related to the underlying physical phenomena. In practice, Gaussian random processes are of great importance in this field.

A scalar random process S_t is said to be Gaussian if the random vector $\{X_{t_1}(\omega), \cdots, X_{t_N}(\omega)\}$ is a Gaussian vector for any finite set of instants $0 \le t_1 < \cdots < t_N \le T$. It is completely defined by prescribing its mean value $\mu_S(t)$ and standard deviation $\sigma_S(t)$ at each time instant, as well as its autocorrelation coefficient function $\rho_S(t_1, t_2)$. Classic forms of autocorrelation coefficient functions are the exponential type ($\exp\left[-|t_1 - t_2|/\tau_S\right]$), the square-exponential type ($\exp\left[-\left((t_1 - t_2)/\tau_S\right)^2\right]$) and the cardinal sine type ($\sin\left[(t_1 - t_2)/\tau_S\right]/\left[(t_1 - t_2)/\tau_S\right]$). Once the process is defined through these properties, trajectories may be simulated for computational purposes by various methods (Fourier decomposition, Karhunen–Loève expansion, EOLE decomposition, etc. [PRE 94], [SUD 07]).

10.2.3. *Poisson and rectangular wave renewal processes*

Point processes appear in numerous situations when similar events occur randomly in time (computer connections to a server, customers arriving at a booth, etc.). In structural reliability problems, they allow the analyst crossings through a limit state surface.

Let us denote by $T_i(\omega), i \ge 1$ the time of the i-th occurrence of an event under consideration (with values in $]0, +\infty[$). The counting function $N_t(\omega)$ is defined by:

$$N_t(\omega) = \sup\{n : T_n(\omega) \le t\} \qquad [10.5]$$

This is a random process whose trajectories are piecewise constant and take integer values, with discontinuities at the time instants where there is an occurrence of the observed phenomenon. Such a process is a Poisson process if it satisfies the following properties:

– for any finite set of instants $0 \le t_1 < \cdots < t_N$, random variables $N_{t_1}, N_{t_2} - N_{t_1}, \cdots, N_{t_N} - N_{t_{N-1}}$ are independent (assuming $N_0 = 0$);

– $\forall\, 0 \leq s < t$,   random   variable $N_t - N_s$ follows   a   Poisson   distribution   with parameter $\lambda(t-s)$, where λ is called *process intensity*. Thus:

$$\mathbb{P}\left(N_t = n\right) = e^{-\lambda t}\frac{\left(\lambda t\right)^n}{n!} \qquad\qquad [10.6]$$

For such processes, it may be proven that the time to first occurrence has an exponential   distribution   with   parameter λ,   (i.e. $\mathbb{P}\left(T_1 \leq t\right) = 1 - e^{-\lambda t}$).   The   time between   two   successive   occurrences $T_{n+1} - T_n$ also   follows   an   exponential distribution.

Poisson processes are useful for constructing rectangular renewal wave processes that are piecewise constant (e.g. exploitation or traffic loads) while changing their amplitude at random time instants. Such processes may be used to model traffic or exploitation loads.

Such a process is defined by (a) the probability density function of the load amplitude (thus of the "jumps" in between) and (b) the Poisson process intensity. A trajectory is depicted in Figure 10.1.

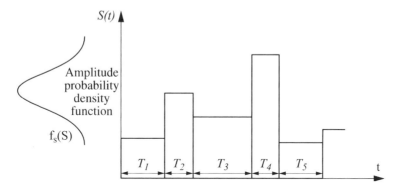

Figure 10.1. *Example of trajectory of a rectangular renewal wave process*

Rectangular renewal wave and Gaussian processes, as well as those obtained by simple transforms (such as lognormal processes obtained by exponentiation of Gaussian processes), allow the analyst to model a large variety of loads for practical applications.

Finally, note that the parameters that define the processes (e.g. mean value) may be random variables as well. This happens for instance in offshore engineering when the environmental loads (wave height) are modeled for different sea states which also occur with some randomness in large time scales.

10.3. Time-variant reliability problems

10.3.1. *Problem statement*

As for time-invariant reliability problems, we assume now that the failure of the structure under consideration is characterized by a limit state function, which may depend on time in two ways: either time may be an input parameter of the function or there are some random processes in its definition (the latter being stationary or non-stationary with time-dependent hyperparameters). Let us denote this limit state function by $g(\boldsymbol{R}(\omega), \boldsymbol{S}_t(\omega), t)$, where $\boldsymbol{R}(\omega) = \{R_1(\omega), \cdots, R_p(\omega)\}^{\mathrm{T}}$ is a random vector, and $\boldsymbol{S}_t(\omega) = \{S_1(\omega), \cdots, S_q(\omega)\}^{\mathrm{T}}$) is a set of scalar random processes, with prescribed joint probability density function.

The main difference between a time-invariant and a time-variant problem lies in the fact that the time instant when failure occurs is not known in the latter case. This instant is the smallest $T \in [0, T]$ such that the limit state function takes a negative value. This leads to the definition of the *cumulative probability of failure*:

$$P_f(0, T) = \mathbb{P}\left(\exists t \in [0, T] : g(\boldsymbol{R}(\omega), \boldsymbol{S}_t(\omega), t) \leq 0\right) \qquad [10.7]$$

In the general case this quantity should not be confused with the *instantaneous probability of failure* denoted by $P_{f,i}(t)$ and defined as:

$$P_{f,i}(t) = \mathbb{P}\left(g(\boldsymbol{R}(\omega), \boldsymbol{S}_t(\omega), t) \leq 0\right) \qquad [10.8]$$

The latter quantity, which could be computed by "freezing" time in the limit state function and using classic methods (Monte Carlo simulation, FORM/SORM, importance sampling, etc.), does not have any particular interpretation, except for *right-boundary problems* which are defined in the next section. In particular, the following inequality holds:

$$P_f(0, T) \geq \max_{t \in [0, T]} P_{f,i}(t) \qquad [10.9]$$

This lower bound is usually very poor and there is little interest in its computation.

10.3.2. *Right-boundary problems*

As noted in the introduction, the degradation of material properties introduces some time dependence into reliability problems. By definition however, this degradation tends to *decrease* the material resistance so that a limit state function of type "*R - S*" decreases monotonically in time. A reliability problem in which all the trajectories of the limit state function are monotonically decreasing is called a *right-boundary problem.* In this specific case only, one can prove that the cumulative failure probability is equal to the instantaneous failure probability computed at the right-boundary of the time interval (and hence the name):

$$P_f(0,T) = P_{f,i}(T)$$

[10.10]

thus solving the time-variant reliability problem reduces to a solving *time-invariant* problem in this case, possibly for various values of T. Classic methods such as FORM/SORM and Monte Carlo simulation may be directly applied.

As an example, consider a steel rebar in a reinforced concrete structure which corrodes in time under the effect of concrete carbonation and/or chloride ingress. The uncorroded rebar cross section $\phi(t)$ may be modeled by:

$$\phi(t) = \begin{cases} \phi_0 & \text{if } t \le T_{init} \\ \phi_0 - 2i_{corr}\kappa(t - T_{init}) & \text{if } t > T_{init} \end{cases}$$

[10.11]

where ϕ_0 is the initial rebar diameter, T_{init} is the initiation time for corrosion (i.e. the time required for the carbonated layer to attain the rebar), i_{corr} is the corrosion current density and κ is a constant.

The performance of the concrete structure may be related to the uncorroded rebar cross section $\phi(t)$: indeed the corroded external layer loses its mechanical resistance and the resulting rust tends to expand into the concrete pores and to crack and shatter the concrete surface ("spalling"). Thus the failure with respect to spalling may be defined by an inequality: $\phi(t) \le (1 - \lambda)\phi_0$ where $\lambda = 0.05$ is a typical value for a service limit state. In this setting, the associated limit state function may be cast as $g(t) = \lambda\phi_0 - 2i_{corr}\kappa(t - T_{init})$, which is clearly decreasing monotonically in time for any realization of the (positive in nature) random variables $\{\phi_0, i_{corr}, T_{init}\}$.

10.3.3. *General case*

As already mentioned, the unique feature of time-variant reliability analysis lies in the fact that the time-to-failure $T(\omega)$ is random and not known in advance: depending on the realizations of the random processes S_t, failure may happen more or less early. This time-to-failure $T(\omega)$ satisfies:

$$\mathbb{P}(T \leq t) = \mathbb{P}\left(\exists \tau \in [0,t] \ : \ g\left(\boldsymbol{R}(\omega), \boldsymbol{S}_\tau(\omega), \tau\right) \leq 0\right) \overset{\text{def}}{=} P_f(0,t) \qquad [10.12]$$

From the above equation it is clear that the cumulative probability of failure $P_f(0,t)$ is nothing but the cumulative distribution function (CDF) of the time-to-failure $T(\omega)$, i.e. the time required for the structure to "cross" the limit state surface. Computing this quantity relies on the evaluation of the *mean outcrossing rate* which is defined in the next section.

10.3.3.1. *Outcrossing rate*

Let us denote by $N_t^+(\omega)$ the number of outcrossings of the zero-level by the limit state function (i.e. when the structure passes from the safe domain to the failure domain) within the time interval $[0,t]$. Failure occurs within this time interval either if it occurs at the initial instant $t = 0$ or if there is *at least* one crossing of the zero-value by the limit state function before time instant t. Thus:

$$P_f(0,t) = \mathbb{P}\left(\left\{g\left(\boldsymbol{R}(\omega), \boldsymbol{S}_0(\omega), t = 0\right) \leq 0\right\} \cup \left\{N_t^+ > 0\right\}\right) \qquad [10.13]$$

After some derivations one can prove that the right-hand side of the expression may be upper bounded as follows [DIT 96], [SUD 07]:

$$P_f(0,t) \leq P_{f,i}(0) + \mathrm{E}\left[N_t^+\right] \qquad [10.14]$$

where $\mathrm{E}\left[N_t^+\right]$ stands for the expected number of outcrossings within $[0,t]$. The outcrossing rate $v^+(t)$ is defined by:

$$v^+(t) = \lim_{h \to 0^+} \frac{\mathbb{P}\left(N^+(t,t+h) = 1\right)}{h} \quad \text{where} \quad N^+(t,t+h) = N_{t+h}^+ - N_t^+ \qquad [10.15]$$

This quantity corresponds to the probability of having exactly one outcrossing in the infinitesimal interval $]t,t+h]$, divided by h. We also consider that the stochastic processes involved in the calculation are regular so that $\lim_{h\to 0^+} \dfrac{\mathbb{P}\left(N^+\left(t,t+h\right)>1\right)}{h} = 0$.

Under this regularity condition, and due to the additivity property of the counting variable in time, we prove that:

$$E\left[N_t^+\right] = \int_0^t v^+(\tau)\,d\tau$$

[10.16]

By substituting [10.16] into [10.14] and recalling [10.9], we can finally obtain the following bounds on the cumulative failure probability:

$$\max_{t\in[0,T]} P_{f,i}(t) \le P_f(0,T) \le P_{f,i}(t=0) + \int_0^T v^+(\tau)d\tau$$

[10.17]

thus solving a time-variant reliability problem (or at least obtaining an upper bound to $P_f(0,T)$) "reduces" to computing the outcrossing rate. Some important analytical results related to simplified problems are now presented, which are used in the following as basic ingredients to solve general problems.

Stationary time-variant reliability problems correspond to cases when the limit state function does not depend explicitly on time and the input random processes (gathered in $S_t(\omega)$) are stationary. The limit state function is formally denoted by $g\left(R(\omega),S_t(\omega)\right)$. In this specific case the outcrossing rate does not depend on time and may be evaluated at any time instant (e.g. $t=0$). Equation [10.17] reduces to:

$$P_{f,i}(t=0) \le P_f(0,T) \le P_{f,i}(t=0) + v^+\,T$$

[10.18]

NOTE: In the case when the limit state function does not depend on random variables but only on stationary random processes (which is formally denoted by $g\left(S_t(\omega)\right)$) the number of outcrossings $N_t^+(\omega)$ is a Poisson process of constant intensity v^+. In that case, a result that is better than the above upper bound is available, namely $P_f(0,T) \equiv F_T(T) \approx 1 - e^{-v^+T}$. However this approximation is no longer valid when g also depends on random variables $R(\omega)$ since the outcrossings no longer occur independently in time ($N_t^+(\omega)$ is not a Poisson process anymore).

The correct estimation is $P_f(0,T) \approx E_R\left[1-e^{-(v^+|R)T}\right]$ in this case and may be computed by specific methods. In the latter equation $v^+|R$ is the *conditional* outcrossing rate and $E_R[.]$ denotes the expectation with respect to these variables; see details in [RAC 98], [SCH 91].

Computing the outcrossing rate of a scalar (vector, respectively) random process through a given threshold (a given hypersurface, respectively) is a complex matter and beyond the scope of this chapter. Readers are referred to [DIT 96], [RAC 04] for a complete treatment. In this chapter we limit the presentation to the classic *Rice's formula*, [RIC 44] which serves as a basis of more advanced results.

Let $S_t(\omega)$ be a scalar differentiable random process and $\dot{S}_t(\omega)$ its derivative process. We can denote their joint probability density function by $f_{S\dot{S}}(s,\dot{s})$. Of interest is the outcrossing rate $v^+(t)$ of this process through a (possibly varying in time) threshold denoted by $a(t)$. Rice's formula reads:

$$v^+(t) = \int_{\dot{a}(t)}^{\infty} (\dot{s}-\dot{a}(t))\, f_{S\dot{S}}(a(t),\dot{s})\, d\dot{s} \qquad [10.19]$$

In case of a stationary random process and a constant threshold (say $a = 0$ in the case of a limit state function for reliability analysis), the above formula reduces to $v^+ = \int_0^\infty \dot{s}\, f_{S\dot{S}}(a,\dot{s})\, d\dot{s}$. As an example, if S_t is a stationary Gaussian process with mean value μ_S and standard deviation σ_S, we can prove that the outcrossing rate for a threshold a is $v_a^{+,Gaussian} = \dfrac{1}{\sqrt{2\pi}}\dfrac{\sigma_{\dot{S}}}{\sigma_S}\varphi\left(\dfrac{a-\mu_S}{\sigma_S}\right)$ where $\varphi(x) = e^{-x^2/2}/\sqrt{2\pi}$ denotes the standard normal PDF. If the Gaussian process is not stationary and if the threshold $a(t)$ is time-dependent, then the outcrossing rate is equal to

$$v^{+,Gaussian}(t) = \dfrac{\sigma_{\dot{S}}}{\sigma_S}\varphi(a(t))\,\Psi\left(\dfrac{\dot{a}(t)}{\sigma_{\dot{S}}/\sigma_S}\right), \text{ where } \Psi(x) = \varphi(x) - x\int_{-\infty}^{-x}\varphi(u)du \quad [CRA67].$$

The calculation of outcrossing rates of vector processes through hypersurfaces makes use of the so-called Belayev's formula, which is presented in [DIT 96], [RAC 04].

10.4. PHI2 method

In the previous section, the basic concepts that are useful for posing and solving time-variant reliability problems have been introduced, namely random processes, outcrossing rate, the cumulative probability of failure and its associated bounds. In order to evaluate equation [10.17] in practice, the outcrossing rate of the limit state function through the zero-level must be computed. As already mentioned, analytical results are available only in very specific cases. Otherwise the analyst has to resort to numerical methods.

Two classes of approaches are well established today in order to solve time-variant reliability problems:

– the so-called *asymptotic* method developed by Rackwitz and co-authors, which estimates the outcrossing rate and its time integral from Rice's formula and various asymptotic approximations, such as the Laplace integration (see [RAC 98], [RAC 04] for details);

– the so-called *PHI2 method*, which is based on solving a system reliability problem and which has been developed in [AND 02], [AND 04], [SUD 08], based on similar work by [HAG 92], [LI 95]. As is explained below, this approach allows us to solve time-variant problems using only the tools available for solving *time-invariant* problems, namely the First Order Reliability Method (FORM) for systems. Thus it may be applied using classic reliability software such as PhimecaSoft [LEM 06] or Open TURNS (www.openturns.org).

By definition, the outcrossing rate is computed from the probability of having one crossing of the limit state surface (zero-level of the limit state function) within two neighbor instants t and $t+h$ (equation [10.15]). In the reliability context, such a crossing means that the structure was in the safe domain at time instant t and in the failure domain at time instant $t+h$. Thus the outcrossing rate may be evaluated as follows (the notation $X_t(\omega) = \{R(\omega), S_t(\omega)\}^\mathsf{T}$ is introduced for the sake of clarity):

$$\nu^+(t) = \lim_{h \to 0^+} \frac{\mathbb{P}\left(\{g(X_t(\omega), t) > 0\} \cap \{g(X_{t+h}(\omega), t+h) \le 0\}\right)}{h} \qquad [10.20]$$

The numerator of the above equation is nothing but the probability of failure of a two-component parallel system which may be estimated by the FORM method for systems ([LEM 09], Chapter 9).

Each component-reliability problem (i.e. at time instants t and $t+h$) is first solved using FORM. Let us denote by $\beta(t)$ and $\alpha(t)$ ($\beta(t+h)$ and $\alpha(t+h)$

respectively) the reliability index and the unit normal vector at the design point that are related to the limit state function $\{g(X_t(\omega),t)\leq 0\}$ ($\{g(X_{t+h}(\omega),t+h)\leq 0\}$ respectively).

The system probability of failure may be computed within the first order approximation by:

$$\mathbb{P}_{\text{FORM}}\left(\{g(X_t(\omega),t)>0\}\cap\{g(X_{t+h}(\omega),t+h)\leq 0\}\right)$$
$$= \Phi_2\left(\beta(t),-\beta(t+h),\alpha(t)\cdot\alpha(t+h)\right)$$

[10.21]

where $\Phi_2(x,y,\rho) = \dfrac{1}{2\pi\sqrt{1-\rho^2}}\displaystyle\int_{-\infty}^{x}\int_{-\infty}^{y}\exp\left[-\dfrac{x^2+y^2-2\rho xy}{2(1-\rho^2)}\right]dx\,dy$ denotes the

cumulative distribution function of the binormal distribution. By combining [10.20] and [10.21], we can prove that the outcrossing rate reads [SUD 08]:

$$v^+(t) = \|\dot{\alpha}(t)\|\,\varphi(\beta(t))\,\Psi\left(\frac{\dot{\beta}(t)}{\|\dot{\alpha}(t)\|}\right) \quad \text{where} \quad \Psi(x) = \varphi(x)-x\int_{-\infty}^{-x}\varphi(u)du \quad [10.22]$$

For stationary time-variant problems, the outcrossing rate does not depend on time and thus simplifies into:

$$v^+ = \frac{\varphi(\beta)}{\sqrt{2\pi}}\|\dot{\alpha}(t)\|$$

[10.23]

We can note the similarity between the two above equations and those given at the end of section 10.3 for the application of Rice's formula to Gaussian processes. In order to give an interpretation of [10.22], we can consider that the FORM method consists of "scalarizing" the outcrossing problem by considering the limit state function as an equivalent scalar process, whose outcrossing of the threshold $\beta(t)$ is of interest.

10.4.1. *Implementation of the PHI2 method – stationary case*

In case of a stationary problem the outcrossing rate is constant in time. It may be computed from equation [10.23] by approximating it by a finite difference scheme:

$$v_{num}^{+} = \frac{\varphi(\beta)}{\sqrt{2\pi}} \left\| \frac{\alpha(t+\Delta t) - \alpha(t)}{\Delta t} \right\|$$ [10.24]

To do so a sufficiently small time increment Δt should be selected. The rule of thumb $\Delta t \approx 10^{-3} \lambda_{min}$ has proved to be efficient and accurate in applications. In this equation, λ_{min} is the smallest correlation length among all the random processes S_t involved in the limit state function [SUD 08]. The various steps for evaluating [10.24] are now summarized:

– the Gaussian vectors $S^1(\omega)$ and $S^2(\omega)$ corresponding to the Gaussian process $S_t(\omega)$ at time instants t and $t+\Delta t$ are first defined. The components S_j^1 and S_j^2 are correlated pairwise with correlation coefficient $\rho_{S_j}(t, t+\Delta t)$, where ρ_{S_j} is the autocorrelation coefficient function of $S_{t,j}$ (equation [10.2]). Note that if the components S_j of S_t are correlated, this so-called *cross-correlation* has to be taken into account as well;

– the "instantaneous" limit state function $g_1(R(\omega), S^1(\omega))$ is defined at time instant t by replacing the random processes S_t by vector S^1 in $g(R(\omega), S_t(\omega))$ and FORM is applied, which yields the reliability index $\beta^{(1)}$ and the unit normal vector $\alpha^{(1)}$;

– the "instantaneous" limit state function $g_2(R(\omega), S^2(\omega))$ is defined at time instant $t+\Delta t$ by replacing the random processes $S_{t+\Delta t}$ by vector S^2 in $g(R(\omega), S_t(\omega))$ and FORM is applied, which yields the reliability index $\beta^{(2)}$ and the unit normal vector $\alpha^{(2)}$;

– from these results, the outcrossing rate [10.24] is evaluated, then the cumulative failure probability is calculated:

$$v_{num}^{+} = \frac{\varphi(\beta^{(1)})}{\sqrt{2\pi}} \left\| \frac{\alpha^{(2)} - \alpha^{(1)}}{\Delta t} \right\| \qquad P_f(0,T) \leq \Phi(-\beta^{(1)}) + v_{num}^{+} \cdot T$$ [10.25]

It is clear that the upper bound linearly increases with T. In order to interpret the result conveniently, the upper bound may be transformed into a "generalized reliability index" $\beta^{inf}(0,T) = -\Phi^{-1}(\Phi(-\beta^{(1)}) + v^{+} \cdot T)$. From the relationship between the probability of failure and the reliability index, the above value is a lower bound to the reliability index, hence the notation β^{inf}. The upper bound reliability index associated to the lower bound in equation [10.17] is simply equal to $\beta^{(1)}$.

Note that two different correlation coefficients are used in the analysis, which should not be confused: the first is the autocorrelation coefficient of each input random process denoted by $\rho_{S_j}(t, t+\Delta t)$; the second is the correlation between the linearized limit state surfaces at time instants t and $t+\Delta t$, which is given by the scalar product $\boldsymbol{\alpha}^{(1)} \cdot \boldsymbol{\alpha}^{(2)}$.

10.4.2. Implementation of the PHI2 method – non-stationary case

In this case, the limit state function explicitly depends on time and/or the input random processes S_t show non-stationarity. Thus the outcrossing rate is evolving in time and should be computed at different time instants, then integrated over $[0,T]$ (equation [10.17]) in order to get the upper bound to $P_f(0,t)$. In practice the time interval is discretized, say $\{t_i = iT/N, i = 0, \cdots, N\}$ and the procedure described in section 10.4.2 is applied at each time instant. The upper bound to $P_f(0,t)$ may be computed using the trapezoidal rule:

$$P_f(0,T) \leq P_{f,i}(0) + \frac{T}{N}\left(\frac{v^+(0)+v^+(T)}{2} + \sum_{i=1}^{N-1} v^+(t_i)\right) \qquad [10.26]$$

Note that the time increment T/N used to compute the integral is not of the same order of magnitude as the time increment Δt which is used for evaluating the outcrossing rate.

10.4.3. Semi-analytical example

Let us consider a cantilever beam of length L and flexural modulus EI that is submitted to a pinpoint load F at its free extremity. The maximum deflection of the beam under quasi-static conditions is equal to $\delta = \dfrac{FL^3}{3EI}$ (the variation of the load in time is assumed slow enough so as to ignore dynamic effects). Of interest is the reliability of the beam with respect to an admissible threshold $\delta_{\max}$ for the maximal deflection. The flexural modulus is supposed to be lognormally distributed (parameters $(\lambda_{EI}, \varsigma_{EI})$). It is also supposed that the logarithm of the load is a stationary Gaussian process S_s of mean value λ_F, standard deviation ς_F and autocorrelation coefficient function $\rho_F(t) = e^{-(t/\tau_F)^2}$, where τ_F is the correlation length. To be able to perform analytical derivations, the limit state function associated with the criterion "the maximal deflection is below the admissible threshold" may be cast as:

$$g(EI, S_t) \overset{def}{=} \ln \delta_{max} - \ln \delta = \ln \delta_{max} - S_t - \ln \frac{L^3}{3} + \ln EI \qquad [10.27]$$

Let us select a particular instant t_0. Random variable $\ln EI$ is Gaussian by definition and may be cast as follows: $\ln EI = \lambda_{EI} + \varsigma_{EI} U_1$, where $U_1 \sim N(0,1)$ is a standard normal variable. Similarly, S_{t_0} is a Gaussian variable of parameters (λ_F, ς_F) that may be cast as $S_{t_0} = \lambda_F + \varsigma_F U_2$, where $U_2 \sim N(0,1)$. After substituting for these expressions in [10.27], the limit state function is revealed to be linear in the reduced variables U_1, U_2. FORM is exact in this case and the associated reliability index reads:

$$\beta^{(1)} = \frac{\ln \delta_{max} - \ln\left(L^3/3\right) - \lambda_F + \lambda_{EI}}{\sqrt{\varsigma_F^2 + \varsigma_{EI}^2}} \qquad [10.28]$$

The coordinates of the unit normal vector to the limit state surface at the design point reads: $\boldsymbol{\alpha}^{(1)} = \left\{ -\varsigma_{EI} / \sqrt{\varsigma_F^2 + \varsigma_{EI}^2}, \varsigma_F / \sqrt{\varsigma_F^2 + \varsigma_{EI}^2} \right\}^T$.

In order to "freeze" the limit state function [10.27] at time instant $t_0 + \Delta t$, we should notice that S_{t_0} and $S_{t_0 + \Delta t}$ are correlated Gaussian variates with correlation coefficient $\tilde{\rho} = e^{-(\Delta t/\tau_F)^2}$ (this number depends on the user choice of $\Delta t \ll \tau_F$, e.g. $\Delta t = 10^{-3} \tau_F$ as suggested above). The isoprobabilistic transform required by FORM in order to handle dependent Gaussian variates leads to the introduction of $U_3 \sim N(0,1)$ and reads (after using the Cholesky decomposition of the correlation matrix):

$$S_{t_0} = \lambda_F + \varsigma_F U_2 \quad , \quad S_{t_0 + \Delta t} = \lambda_F + \varsigma_F \left(\tilde{\rho} U_2 + \sqrt{1 - \tilde{\rho}^2} U_3 \right) \qquad [10.29]$$

The instantaneous limit state function at time instant $t_0 + \Delta t$ is revealed to be linear in the three reduced variables U_1, U_2, U_3. The (exact) reliability index is identical to that obtained at time instant t (equation [10.28]), which is logical since the problem is stationary. The unit normal vector now reads:

$$\alpha^{(2)} = \left\{ -\varsigma_{EI} \Big/ \sqrt{\varsigma_F^2 + \varsigma_{EI}^2} \, , \, \tilde{\rho}\varsigma_F \Big/ \sqrt{\varsigma_F^2 + \varsigma_{EI}^2} \, , \, \sqrt{1-\tilde{\rho}^2}\varsigma_F \Big/ \sqrt{\varsigma_F^2 + \varsigma_{EI}^2} \right\}^{\mathrm{T}}$$

In order to finish the computation, numerical values should be given to the various parameters. Then $\beta^{(1)} = \beta^{(2)}$ is computed from equation [10.28]. The $\alpha^{(i)}$-vectors are evaluated and the values are used to compute the outcrossing rate and the probability of failure using [10.25].

10.5. Industrial application: truss structure under time-varying loads

Consider the elastic 23-bar truss depicted in Figure 10.2 that has already been presented in Chapter 8.

Of interest is the time-variant reliability of such a truss structure under time-varying loads applied on the upper part.

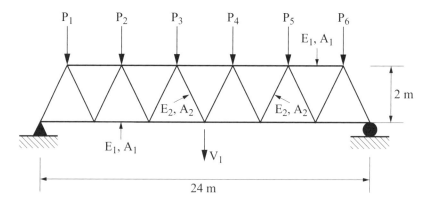

Figure 10.2. *23-bar truss structure*

The input random variables are described in Table 10.1. The six vertical loads are modeled by a *single* stationary Gaussian process P_t with mean value 50 kN, standard deviation 7.5 kN, and Gaussian autocorrelation coefficient function

$\rho_P(t) = e^{-(t/\tau_P)^2}$ where the correlation length is $\tau_P = 1$ day . According to this value the time variation of the load is sufficiently slow so that inertial effects can be neglected: the quasi-static solution is thus valid. The time-variant reliability of the truss with respect to an admissible maximal deflection reads:

$$g(E_1, A_1, E_2, A_2, P_t) = v_{max} - |\mathcal{M}(E_1, A_1, E_2, A_2, P_t)| \le 0, \; v_{max} = 16 \text{ cm} \qquad [10.30]$$

where $\mathcal{M}(E_1, A_1, E_2, A_2, P_t)$ is the maximal deflection computed by finite element analysis. Due to stationarity, a single evaluation of the outcrossing rate is necessary.

The initial problem has 4 basic random variables and a single random process. Using the PHI2 method, it is transformed into two (time-invariant) FORM analyses which involve 4+2 = 6 random variables (including one for P_t and one for $P_{t+\Delta t}$). The time increment is $\Delta t = 10^{-3}$.

Random variable	Distribution	Mean value	Standard deviation
E_1, E_2 (MPa)	LogNormal	210,000	21,000
A_1 (cm²)	LogNormal	20	2
A_2 (cm²)	LogNormal	10	1
P_t (kN)	Gaussian process	50	7.5

Table 10.1. *23-bar truss: description of the random variables*

The instantaneous reliability analysis yields β = 4.032 and $\alpha^{(1)}$ = {-0.533447, -0.067651, -0.533447, -0.067651, 0.649397, 0.}^{T}. At time instant $t + \Delta t$ the same reliability index is obtained and the unit normal vector is $\alpha^{(2)}$ = {-0.533447, -0.067651, -0.533447, -0.067651, 0.649396 0.000918}^{T}.

It may be observed that only the last two components of the α-vector (*i.e.* the ones related to the random process) change between the two time instances. Using equation [10.18] yields the outcrossing rate $v^+ = 4.3.10^{-5}$ /day. The upper bound to the cumulative failure probability is obtained from [10.25].

The evolution in time of this quantity is plotted in Figure 10.3. These results show that the cumulative failure probability may be greater than the instantaneous probability of failure by orders of magnitude. Note that the latter corresponds to the time-invariant case when the loads are modeled by a single random variable.

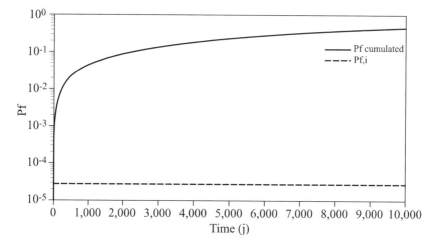

Figure 10.3. *23-bar truss: cumulative probability of failure*

10.6. Conclusion

Structural reliability methods are now well established for time-invariant problems and they are used on a regular basis in industrial applications, as shown throughout this book. Time-variant reliability analysis is far less mature. First, the handling of random processes instead of random variables introduces some additional abstract concepts. Moreover, the quantity that is of interest, namely the *cumulative* probability of failure, is rather difficult to compute.

In this chapter, only the basic concepts have been introduced. Stochastic dynamics problems, in particular, have not been addressed. Specific methods have been introduced for solving such problems, see e.g. [KRE 83], [LUT 03], [SOI 01]. Only the PHI2 method has been presented in detail: it allows the analyst to compute the outcrossing rate using FORM for systems. This means that only classical *time-invariant* tools may be used for solving time-variant problems, which are available within many reliability softwares.

Finally, note that the Monte Carlo method has not been presented here in the context of time-variant problems. Its use would require sampling trajectories of random processes and then the solution of transient mechanical problems. This is obviously a very tedious and costly approach that should be used only as a last resort, especially for nonlinear dynamics treated in the time domain (e.g. for seismic analysis of structures).

10.7. Bibliography

[AND 02] ANDRIEU-RENAUD C., Fiabilité mécanique des structures soumises à des phénomènes physiques dépendant du temps, PhD thesis, Blaise Pascal University, Clermont-Ferrand, France, 2002.

[AND 04] ANDRIEU-RENAUD C., SUDRET B., LEMAIRE M., "The PHI2 method: a way to compute time-variant reliability", *Reliab. Eng. Sys. Safety*, vol. 84, p. 75-86, 2004.

[CRA 67] CRAMER H., LEADBETTER M., *Stationary and Related Processes*, John Wiley & Sons, Chichester, 1967.

[DIT 96] DITLEVSEN O., MADSEN H., *Structural Reliability Methods*, John Wiley & Sons, Chichester, 1996.

[HAG 92] HAGEN O., "Threshold up-crossing by second order methods", *Prob. Eng. Mech.*, vol. 7, p. 235-241, 1992.

[KRE 83] KREE P., SOIZE C., *Mécanique aléatoire*, Dunod, Paris, France, 1983.

[LEM 09] LEMAIRE M., *Structural reliability*, ISTE Ltd, London and John Wiley & Sons, New York, 2009.

[LEM 06] LEMAIRE M., PENDOLA M., "PHIMECA-SOFT", *Structural Safety*, vol. 28, p. 130-149, 2006.

[LI 95] LI C., DER KIUREGHIAN A., "Mean out-crossing rate of nonlinear response to stochastic input", in LEMAIRE M., FAVRE J., MEBARKI A. (eds), *Proc. 7th Int. Conf. on Applications of Stat. and Prob. in Civil Engineering (ICASP7)*, Paris, p. 1135-1141, Balkema, Rotterdam, The Netherlands, 1995.

[LIN 67] LIN Y.K., *Probabilistic Theory of Structural Dynamics,* McGraw–Hill, New York, USA, 1967.

[LUT 03] LUTES L., SARKANI S., *Random Vibrations: Analysis of Structural and Mechanical Systems*, Butterworth–Heinemann, Oxford, 2003.

[MEL 99] MELCHERS R., *Structural Reliability Analysis and Prediction*, John Wiley & Sons, Chichester, 1999.

[PRE 94] PREUMONT A., *Random Vibrations and Spectral Analysis*, Kluwer Acamedic, Dordrecht, The Netherlands, 1994.

[RAC 98] RACKWITZ R., "Computational techniques in stationary and non-stationary load combination – A review and some extensions", *J. Struct. Eng.*, vol. 25(1), p. 1-20, 1998.

[RAC 01] RACKWITZ R., "Reliability analysis – A review and some perspectives", *Structural Safety*, vol. 23, p. 365-395, 2001.

[RAC 04] RACKWITZ R., Zuverlässigkeit und Lasten im konstruktiven Ingenieurbau, Lecture notes, Technical University of Munich, Germany, 2004.

[RIC 44] RICE S., "Mathematical analysis of random noise", *Bell System Tech. J.*, vol. 23, p. 282-332, 1944.

[SOI 01] SOIZE C., *Dynamique des structures – Eléments de base et concepts fondamentaux*, Ellipses, Paris, France, 2001.

[SCH 91] SCHALL G., FABER M., RACKWITZ R., "The ergodicity assumption for sea states in the reliability assessment of offshore structures", *J. Offshore Mech. Arctic Eng.*, ASME, vol. 113(3), p. 241-246, 1991.

[SUD 07] SUDRET B., Uncertainty propagation and sensitivity analysis in mechanical models – Contributions to structural reliability and stochastic spectral methods, Thesis, Blaise Pascal University, Clermont-Ferrand, France, 2007.

[SUD 08] SUDRET B., "Analytical derivation of the outcrossing rate in time-variant reliability problems", *Struc. Infra. Eng.*, vol. 4(5), p. 353-362, 2008.

Chapter 11

Bayesian Inference and
Markov Chain Monte Carlo Methods

11.1. Introduction

Many different types of failure data appear in civil engineering studies. In the specific case of lifetime analysis, data are often censored (see Chapter 4) and thus are not very informative. When data does not provide much information, for whatever reason, it is said that the data are poor. It appears that Bayesian inference through Markov Chain Monte Carlo (MCMC) simulation methods could be useful in such cases.

These methods allow Probability Density Functions (PDF) to be simulated by generating a Markov chain whose stationary distribution is the target PDF. A Markov chain is a discrete time random process such that its current value is only dependent on the past through its previous value.

Section 11.2 introduces the Bayesian inference context underlying MCMC methodology. MCMC methods to update input parameters in the setting of poorly informative data sets are presented in section 11.3. Finally, an application to lifetime estimation for an in series system with censored and missing data is presented in section 11.4. Another application of MCMC methods, relating to a case study for warping prediction in nuclear power plant containment components, is presented in Chapter 12.

Chapter written by Gilles CELEUX.

11.2. Bayesian Inference

Bayesian methodology [ROB 06] allows prior knowledge of the parameters of the density of a random vector to be combined with observed values of that vector. Let X be a random vector with PDF $f_X(x;\theta)$ characterized by a hyperparameter vector θ of size n_θ, with *a priori* distribution $p_\theta(\theta)$ defined by $D_\Theta \subset R^{n_\theta}$, and a set of observations $X = \{\xi^{(1)}, \_, \xi^{(Q)}\}$. Using the Bayes theorem, we get a posterior distribution derived from the prior distribution and from the observations density $p_q(q/c)$:

$$p_q(q/c) = \frac{1}{c} p_q(q) L(q;c) \qquad [11.1]$$

where L is the *likelihood function*, defined with independent observations by:

$$L(\theta;\mathcal{X}) = \prod_{q=1}^{Q} f_X\left(x^{(q)};\theta\right) \qquad [11.2]$$

and where c is the normalizing constant defined by $c = \int_{D_\Theta} p_\Theta(\theta) L(\theta;\mathcal{X}) d\theta$.

This leads to the predictive density of the random vector X:

$$f_X^p(x) = \int_{D_\Theta} f_X(x,\theta) f_\Theta(\theta) d\theta, \qquad [11.3]$$

which is also known as the *integrated likelihood* of the model under discussion. This integrated likelihood can be regarded as a Bayesian criterion to select a relevant model (see [ROB 06]).

Several examples will be considered in this chapter:

– the observations $(x_1, ..., x_n)$ are the lifetimes of n pieces of equipment, and a standard distribution to model these lifetimes is the Weibull distribution. It is parameterized with a scale parameter η and a shape parameter β. Its PDF (with $\theta = (\beta, \eta)$ denoting its parameters) is:

$$f(x;\theta) = \beta \left(\frac{x}{\eta}\right)^{\beta-1} e^{-(\frac{x}{\eta})^\beta}$$

– the observations $(x_1,..., x_n)$ are the weights of n manufactured objects, assumed to arise from a Gaussian distribution with mean μ and variance σ^2 with PDF:

$$f(x;\theta) = \frac{1}{\sqrt{2\pi\sigma}} e^{-\frac{1}{2}(\frac{x-\mu}{\sigma})^2} ,$$

and with a vector parameter to be estimated, $\theta = (\mu, \sigma^2)$.

11.2.1. *Bayesian estimation of the mean of a Gaussian distribution*

Let $N(\theta, \sigma^2)$ be the distribution of a Gaussian variable with unknown mean θ and with known variance σ^2. A prior distribution $\Pi() = N(\mu, \tau^2)$ is assumed for θ, μ and τ^2, being the hyperparameters to be fixed by experts (described below). Thus, the posterior distribution of θ knowing the observations x have a Gaussian distribution is:

$$N(\tilde{\theta}, \tilde{\tau}) = N(\frac{\sigma^2\mu + n\tau^2\bar{x}}{\sigma^2 + n\tau^2}, \frac{\sigma^2\tau^2}{\sigma^2 + n\tau^2})$$

where $\bar{x} = \frac{1}{n}\sum_{i=1}^{n}x_i$ is the empirical mean of the data.

An example concerning wood flexion test can be found in section 5.3.2. The empirical mean from six measurements is 147.5 MPa with a standard deviation of 21 Mpa, and the prior distribution was $\theta \sim N(\mu = 140, \tau = 20)$. Thus, the posterior distribution of the mean is $N(\tilde{\theta}_6, \tilde{\tau}_6^2) = N(146.3; 7.9^2)$.

This example illustrates the role and impact of hyperparameters on Bayesian parameter estimation. The Bayesian estimator appears to give a compromise between the empirical mean and the prior parameter guess θ. Thus the Bayesian estimator has a shrinking effect on the empirical estimator. Analyzing the posterior distribution form shows the effect of the prior variance τ^2 on the Bayesian estimator. If τ^2 is chosen too small, the prior distribution will be too great with regard of the observed sample. Typically, τ^2 has to be large to ensure that the effect of the prior distribution is not too great. In the same way, the effect of the sample size n is apparent. The greater n is, the smaller the effect of the prior distribution. And, when n is large, as is desirable, there is little difference between maximum likelihood and Bayesian estimators.

Thus, Bayesian inference is definitively the most natural and efficient approach for a relevant and useful statistical inference with small sample sizes. However, Bayesian inference has some disadvantages that are now considered:

– translating prior information in prior distributions (choosing the form of the prior densities, and especially the values of the hyperparameters, etc.) is a sizeable task;

– what happens when no prior information is available? How can we design non-informative prior distributions?

– deriving the normalizing constant c in [11.1] often involves quite difficult multivariate integration in order to compute the posterior distribution of the parameters to be estimated;

– approximating posterior distributions can be achieved through Monte Carlo methods. But an efficient use of Monte Carlo algorithms involves numerical and computational difficulties. Overcoming those difficulties can be costly and requires very specific skills.

11.3. MCMC methods for weakly informative data

11.3.1. *Weakly informative statistical problems*

The main source of wealth for a statistician is available data. The more complex a probabilistic model is, the more numerous the data that are needed for a relevant statistical analysis. Thus, to assess the validity of a statistical analysis, the sample size n should be compared to the number v of parameters of the model at hand. Typically with a two-parameter probability distribution **R**, difficulties can arise as soon as $n < 20$. Problems become serious when $v \approx n$. These days, statisticians are increasingly facing sample sizes smaller than the dimension p of the data set. Moreover, it can happen that each data brings only partial information about the model. For instance, in reliability, many lifetime data could be censored. Remember that a lifetime datum is right censored in c, if we simply know that $x > c$ (and left censored in c, if we simply know that $x < c$). A lifetime x is interval censored if we simply know that $a < x < b$.

Latent structure models are another example of missing data models. These models are concerned with situations where some label data related to observed data are missing. In reliability, *competing risk* models are an example of latent structure models. Let i be a piece of equipment made of k components $C_1, \ldots, C_k$ arranged in series. The complete data are (x_i, z_i), $i = 1, \ldots, n$, x_i being the lifetime of equipment i and z_i being the label of the component which has caused the equipment failure. But often the z_i's are missing…

Small and poorly informative samples jeopardize statistical analysis in several ways, from the high variability of parameter estimates, the high influence of outliers, or through contrast exaggeration, etc. In what follows, we focus on Bayesian inference which provides a well-grounded and relevant framework to statistical analysis from such samples.

11.3.2. *From prior information to prior distributions*

The first task is to choose the form of prior distribution. It is not a too sensitive task and, as we will see, choosing values for prior hyperparameters is a more sensitive. Typically for a real-value parameter, a prior distribution with two parameters should be chosen. This allows information on the mean value of the parameter given by an expert in reliability to be entered, and the variability of this mean value (given by the expert) to be taken into account. Thus, using two hyperparameters allows the statistician to quantify his confidence level on the prior information provided by the expert in reliability. The statistician has to give a reasonable weight to the prior information with respect to the information provided by observations. In particular, this uncertainty hyperparameter allows the impact of a doubtful expert to be reduced, without biasing his opinion.

Sensitivity analyses are needed to choose the hyperparameters in a proper way to get stable Bayesian estimates which do not give too important a weight to the opinion of the expert in reliability. Those sensitivity analyses are usually done on an empirical basis by trial and error. For many standard models (except Weibull models), conjugate prior distributions are available. These conjugate prior distributions are probability distributions such that the posterior distribution belongs to the same parametric family of distributions as the prior distributions. Readers can find a list of the most used prior conjugate families of distribution in [ROB 06]. Some examples include:

– a univariate Gaussian model, with a Gaussian prior for the mean and inverse gamma prior for the variance;

– a multivariate Gaussian model, with a Gaussian prior for the mean vector and Wishart distribution for the variance matrix;

– an exponential model, with a Gamma prior.

It is important to say that the most delicate and sensitive step is choosing the hyperparameters of the prior distribution. Readers can find prior choices for Weibull lifetime models in [BAC 98] and an example of choosing prior hyperparameters for a simple conjugate prior distribution is given below.

An example of hyperparameter choice: consider a piece of equipment whose lifetime follows an exponential distribution with mean η. The chosen prior distribution is a conjugate gamma distribution $\Pi(\eta) = G(a, b)$, with mean a/b and variance a/b^2. The hyperparameters to be fixed are a and b. First, the ratio a/b is chosen in order to take into account the opinion of expert(s) in reliability on the mean lifetime of this equipment. Then, the statistician's viewpoint is taken to fix a small b value in order to get a prior distribution with a large enough variance. A default choice is to ensure that the prior information does not exceed the information provided by an observed lifetime value. In the present case, this rule leads to us taking $b \leq 1$ (see [BAC 98], for more details).

Non-informative prior distributions are a general and important family of prior distributions. They are useful when no prior information on the process described with the data at hand is available. In such cases, the prior distribution should express our ignorance on the model parameters but it must not depend on the parameterisation. An apparently natural choice consists of assuming a uniform distribution for the parameter. But this is controversial since it depends on the parameterisation. For instance, a "uniform" prior for the dispersion parameter around a mean will be different if we consider that this parameter is the variance σ^2, or the standard deviation σ, or the precision $1/\sigma^2$.

The Jeffreys solution to this invariance problem (see [ROB 06]) consists of considering a distribution proportional to the square root of the Fisher information $I(\theta)$ of the parameter θ:

$$I(q)=-E[\frac{\P^2}{\P q^2}\log(f(x;q))].$$

For a Gaussian distribution $N(\mu,\sigma^2)$, the Jeffreys prior distribution is $\Pi(\theta) \propto 1/\sigma^2$, whilst for a Weibull distribution with shape parameter β and scale parameter η, the Jeffreys prior distribution is $\Pi(\beta, \eta) \propto 1/(\eta\beta)$.

Notice that, in general, Jeffreys priors are not probability distributions. They are said to be *improper* prior distributions. This is not a problem because the associated posterior distribution is a proper probability distribution.

11.3.3. *Approximating a posterior distribution*

The most difficult task is to approximate a posterior distribution given, for example, by [11.1]. Calculating the posterior distribution involves computing a highly multivariate integral as soon as the probabilistic model becomes somewhat complex. This difficulty has limited the impact of Bayesian inference for a long time. But, since the early 1990s, Markov Chain Monte Carlo (MCMC) methods have become a popular and efficient tool to approximate integrals and posterior distributions. Consequently, Bayesian inference has dramatically increased its influence on applications and methodological developments.

Standard numerical integration tools are limited to small dimension settings, while Laplace approximation is essentially efficient to recover posterior distributions close to a multivariate Gaussian distribution. Thus, these deterministic methods are being replaced more and more by Monte Carlo methods making use of random simulations. There are two kinds of Monte Carlo methods: MCMC methods consisting of simulating Markov chains whose stationary distribution is the posterior distribution to be estimated, while "Importance Sampling" methods compare the target distribution to an instrumental distribution which is supposed to be easy to generate and close to the target distribution.

11.3.4. *A popular MCMC method: Gibbs sampling*

The principle of MCMC methods is to simulate a Markov chain whose limit distribution is the posterior distribution at hand. *Gibbs sampling* is certainly the most popular MCMC method and it can be regarded as a specific case of the general Metropolis–Hastings algorithm that we present in the next subsection. Starting from a vector $\theta^0 = (\theta_1^0, \ldots, \theta_v^0)$, Gibbs sampling consists of drawing a new realisation of each vector coordinate at random according to its full conditional posterior distribution knowing the data and the other vector θ coordinates, and repeating those random drawings a large number of times. Thus, the iteration $(i + 1)$ of Gibbs sampling consists of simulating:

– θ_1^{i+1} according to the conditional posterior distribution $\Pi(\theta_1/\theta_2^i, \ldots, \theta_v^i, \boldsymbol{x})$,

– θ_2^{i+1} according to the conditional posterior distribution $\Pi(\theta_2/\theta_1^{i+1}, \theta_3^i, \ldots, \theta_v^i, \boldsymbol{x})$,

– ...

– θ_d^{i+1} according to the conditional posterior distribution $\Pi(\theta_d/\theta_1^{i+1}, \ldots, \theta_{v-1}^{i+1}, \boldsymbol{x})$.

Obviously, a *burn-in* period L is necessary before the generated Markov chain is in its stationary distribution. Then the sequence $(\theta^{L+1}, \ldots, \theta^{L+M})$ is a non independent sample from the limit distribution of the Markov chain and can be regarded as a non independent sample of the posterior distribution $\Pi(\theta/\boldsymbol{x})$ [ROB 04].

In order to get an independent sample from this correlated sequence, only one value among t values is considered (typically $t = 5$ or 10.) Thus, for any function of interest h, the empirical mean $\dfrac{t}{M}\displaystyle\sum_{i=L+1}^{L+M} h(\theta^{L+1+(i-L-1)t})$ tends toward the posterior

expectation $E_\Pi(h(\theta))$ as M tends to infinity.

The reason for the success of Gibbs sampling lies in the fact that, in most situations, simulating a full conditional distribution is easier than simulating from a joint or a marginal distribution. Since each step of Gibbs sampling is simple, it does not require specific technical skill. However, in some cases, there is the need to use a general Metropolis–Hastings algorithm inside a Gibbs sampling iteration to simulate some particular conditional posterior distributions. The Metropolis–Hastings algorithm, which is a more general simulation method than Gibbs sampling, is now presented.

11.3.5. *Metropolis–Hastings algorithm*

The Metropolis–Hastings algorithm is probably the oldest MCMC method (the Metropolis article dates to 1953, whilst the Hastings one dates from 1970). This method has links to the *accept–reject* method for simulating a PDF, and that method is presented first below. Next, a detailed balance condition ensuring that a Markov chain converges to its unique stationary distribution is presented, before the Metropolis–Hastings algorithm itself.

11.3.5.1. *Accept–reject method*

Let $f(x) = g(x)/K$, be a PDF of interest, K being a normalizing constant, possibly unknown. We assume that a PDF $h(x)$, easy to simulate (for instance by the direct inversion method) is available and that there a constant c such that $g(x) \leq c\,h(x)$ for any x also exists. The accept–reject algorithm to simulate a sample from the PDF $f(x)$ is then as follows:

– generate z according to h and u according to a uniform distribution of $[0, 1]$;

– if $u \leq g(z)/ch(z)$, accept z as a realisation from f otherwise go back to the previous step.

It can be easily shown (for example, see [ROB 04]) that the accepted value is actually drawn from PDF f.

This procedure is iterated until the number n of realisations from f is obtained. The value of the constant c is clearly important since the expectation of the acceptation frequency is $1/c$. Thus, choosing h "near" g is desirable.

11.3.5.2. *MCMC and detailed balance condition*

To generate a sample from PDF *f*, MCMC methods consist of generating a Markov chain with stationary distribution *f*. The problem is thus to choose the transition kernel *P(x,dy)* of a Markov chain with stationary distribution *f*. For that purpose, the Metropolis–Hastings algorithm has to check the following detailed balance condition. Assume that the transition kernel can be written as:

$$P(x, dy) = p(x, y) + r(x)\delta_x(dy)$$

with *p(x, x)* = 0, $\delta_x(dy)$ = 1, if *x* belongs to *dy* and 0 otherwise, and where *r(x)* = 1-∫*p(x, y)dy* is the probability that the chain stays in *x*. Then, if the function *p(x, y)* satisfies the following detailed balance condition:

$$f(x)p(x, y) = f(y)p(y, x)$$

where *f* is the unique stationary distribution of the chain with transition kernel *P(x,.)* (see, for example, [ROB 04]).

11.3.5.3. *The Metropolis–Hastings algorithm itself*

We are now in position to describe the Metropolis–Hastings algorithm. Assume that a PDF *q(x, y)* (∫*q(x,y)dy* = 1), is available which is easy to simulate. This PDF can be interpreted as follows: as the generated process is *x*, the next point *y* is generated according to *q(x, y)*. Now, *q(x, y)* does not check (in general) the detailed balance condition. For instance, *(x, y)* exists such that:

$$f(x)q(x, y) > f(y)q(y, x).$$ [11.4]

This inequality means that the process generated allows more moves from *x* to *y* than from *y* to *x*. In order to correct this, a probability *a(x, y)* < 1 to move from *x* to *y* is introduced. If a move from *x* to *y* is not allowed, the chain stays in *x*. Thus, in the Metropolis–Hastings (MH) process, a move from *x* to *y*, *x* ≠ *y* occurs with the probability:

$$p_{MH}(x, y) = q(x, y)a(x, y).$$

Notice that there is a difference here to the accept–reject method which, unlike MH, does not accept to stay in the same position.

Now, the move probability *a* has to be specified. First, to get balanced moves, the reverse probability of move *a(y, x)* has to be as great as possible, that is to say that $a(y, x) = 1$. In addition, a has to be chosen to ensure the detailed balance condition for p_{MH}:

$$f(x)q(x, y)a(x, y) = f(y)q(y, x)a(y, x) = f(y)q(y, x).$$

This leads to $a(x, y) = f(y)q(y, x)/f(x)q(x, y)$. Since inequality [11.4] can be reversed, we finally get:

$$a(x, y) = \min\left(\frac{f(y)q(y, x)}{f(x)q(x, y)}, 1\right), \text{ if } f(x)q(x,y) > 0 \text{ and 1 otherwise.}$$

To terminate the transition kernel definition of the MH chain, the probability $r(x)$ to remain in x has to be given. This is:

$$r(x) = 1 - \int q(x, y)a(x, y)dy.$$

Finally, the transition kernel of the MH chain can be written as:

$$P_{MH}(x, dy) = q(x, y)a(x, y)dy + [1 - \int q(x, y)a(x, y)dy]d_x(dy)$$

and has the desired form. Consequently, it implies that f is the stationary distribution of the MH chain.

Note that:

– since $a(x, y)$ is a ratio not dependent on any normalizing constant, it appears that the Metropolis–Hastings algorithm does not need to know the normalizing constant of the target PDF f;

– if q is symmetric, where $q(x,y)=q(y,x)$ for any (x, y), the Metropolis–Hastings algorithm accepts a move with probability $f(y)/f(x)$;

– the MH chain converges to its limit distribution f if it is irreducible and aperiodic. This is ensured by the fact that $q(x, y)$ is positive on the support of f (namely the set where f is positive).

Choosing a good instrumental density $q(x, y)$ is clearly quite important. Three standard choices are as follows:

– an independent sampler where $q(x, y) = q(y)$ does not depend on x. In such a case, q is commonly chosen as the PDF of a Gaussian distribution. It is often an excellent choice, but it could be a poor choice;

– a random walk which, by its very nature, takes into account the previously simulated value to generate the following value is also a standard choice;

– a strategy inspired by the accept–reject algorithm is choosing a PDF $q(x)$ dominating $h(x) = f(x)K$. In such a case, it is crucial that the PDF q has heavier tails than the target PDF f.

Other choices of the "instrumental" PDF q are described in books dealing with MCMC methods, such as [ROB 04]. Moreover, readers will find an example of the use of a particular Metropolis–Hastings algorithm in the following chapter of this book.

11.3.6. *Assessing the convergence of an MCMC algorithm*

For any MCMC method (Gibbs sampling, Metropolis–Hastings algorithm or more sophisticated samplers), there is the need to assess the convergence of the generated Markov chain. More precisely the user has to answer the following questions:

– how many iterations L are needed to ensure that the chain has converged to its stationary distribution? then,

– how many iterations M are needed to get a good approximation of the target distribution f?

These difficult questions have received a lot of attention and an excellent account of the most useful available methods is presented in [ROB 04] (see also [ELA 06]). However, no reference method supported by strong theoretical results is really available. Thus, the convergence of MCMC chains is often assessed on pragmatic grounds. Here, we confine ourselves to the presentation of simple popular heuristics to assess MCMC convergence:

– it is necessary to run several MCMC chains in parallel. If the numbers L and especially M have been properly chosen, Bayesian inference should not be very sensitive to the selected MCMC chain;

– the burn-in length L is fixed from simple figures displaying the evolution of some parameter values as a function of the number of iterations. The value of L is chosen to ensure that the corresponding values fluctuate symmetrically around their mean;

– the number M has to be large enough to ensure that the MCMC chain has been simulated a sufficient number of times to get a reasonable image of the target posterior distribution. A simple way to proceed is to increase M and to stop when the posterior parameter means become stable.

In any case, it is important to keep in mind that there is a price to be paid when using an MCMC method. It is true that MCMC algorithms avoid difficult calculations, numerical difficulties (degenerate solutions, dividing by zero occurrences, etc.) or painful coding time. But MCMC algorithms can often converge quite slowly and need to be run a tremendous number of iterations to provide good results. Roughly speaking L is about several thousand of iterations and M is about several ten thousand iterations in many cases. Moreover, it is important to keep in mind that assessing the convergence of an MCMC chain is quite a difficult task, especially for complex models with small sample sizes.

11.3.7. *Importance sampling*

Importance sampling is a Monte Carlo simulation tool that differs by its very nature from MCMC methods. Importance sampling aims to approximate integrals of the form:

$$T(\Phi) = \int \Phi(\theta)\pi(\theta / x)d\theta$$

via an instrumental distribution $\rho(\theta)$, by using the identity:

$$T(\Phi) = \int \frac{\Phi(\theta)\pi(\theta / x)}{\rho(\theta)} \rho(\theta)d\theta .$$

More precisely, importance sampling consists of two steps and can be completed with an additional simulation step from the instrumental distribution $\rho(\theta)$:

1. Generate M independent realisations $(\theta^1, ..., \theta^M)$ from the PDF $\rho(\theta)$;

2. Compute the importance weights w_i proportional to $\dfrac{p(q^i / x)}{\rho(q^i)}$ for $i = 1, ..., M$,

and the probabilities $p_i = w_i / \sum_{j=1}^{M} w_j$;

3. Generate $(\tilde{\theta}^1, ..., \tilde{\theta}^L)$ an independent sample from $(\theta^1, ..., \theta^M)$ according to the distribution: $(p_i, i = 1, ..., M)$, $L < M$.

This leads to an approximation of the integral $T(\Phi)$ with the empirical mean $\sum_{i=1}^{M} p_i \Phi(\theta_i)$. This approximation can be regarded as a standard numerical approximation where the integration nodes are the points θ_i with coefficients p_i.

Choosing an instrumental density is obviously crucial to ensure good performances. First, it is important to notice that the convergence of the integral $T(\Phi)$ occurs only if the support of the importance density $\rho(\theta)$ includes the support of the posterior density $\Pi(\theta/x)$ (see, for example, [ROB 04]). Now, the nearer the instrumental density is from the posterior density, the better the importance sampling approximation is. The art of importance sampling is to propose efficient means to design a relevant instrumental density. In general, importance sampling is considered to perform quite well when the instrumental density is chosen properly; however, it can be terrible otherwise. Moreover, it is important to remark that importance sampling often performs poorly in high dimensions. In the case study in the next section, we present an adaptive importance sampling method in a hidden structure model context.

11.4. Estimating a competing risk model from censored and incomplete data

To illustrate Bayesian inference via Monte Carlo methods, an in series component is considered. For instance, this piece of equipment could be part of a pipeline or a drainage element. For simplicity, we will restrict our attention to two-component systems.

The problem is to characterize the equipment's lifetime T. It is assumed that the component lifetimes C_1 and C_2 follow independent Weibull distributions W_1 and W_2 (see Chapter 4) with densities:

$$f_{W_i}(x;\theta_i) = \beta_i \left(\frac{x}{\eta_i}\right)^{\beta_i - 1} e^{-\left(\frac{x}{\eta_i}\right)^{\beta_i}} \quad ; i = 1,2 \qquad [11.5]$$

where θ_i, $i=1, 2$ are the parameters to be estimated, with $\theta_i = (\beta_i; \eta_i)$, β_i and η_i being, respectively, the shape and scale parameters of the Weibull random variable W_i. These parameter have to be estimated from n observed lifetimes $(x_1, ..., x_n)$ of the whole equipment. In the case considered, we have $\beta_1 = 1.5$, $\eta_1 = 100$, $\beta_2 = 2.5$ and $\eta_2 = 120$, with a sample size of $n = 100$, and a fixed censoring time $c = 40$. In those conditions $r = 30$ failure times were observed, 26 caused by the component C_1 and 4 by the component C_2. But the labels of the component causing the failures are unknown.

Thus, the observed data are $(t_1, ..., t_r, c_1, ..., c_{n-r})$, that is, the r failure times smaller than c are ranked in increasing order, and there are $(n - r)$ right censored data. The completed data are $(t_1, e_1), ..., (t_n, e_n)$, where the t_i's are the n failure times, and the e_i's indicate the labels of the component which have caused the failure: $e_i = 1$ or $e_i = 2$ and $e_i = 0$ for a censored time. If the e_i labels were available, it would be simple to estimate the parameters (β_1, η_1) and (β_2, η_2). But the estimation problem becomes difficult when the e_i labels are missing and the number r failure times is low.

This model is a *competing risk model* where the random variable density to be estimated is $T = \min(W_1, W_2)$, which is the minimum of two Weibull random variables. Since the failure labels are unknown, the maximum likelihood could be derived using the *Expectation Maximization* (EM) algorithm [DEM 77], which is a reference algorithm for hidden structure models, or with its stochastic version the *Stochastic Expectation Maximization* (SEM) algorithm. Details of the equations of these two maximum likelihood algorithms are not detailed here but they can be found in [BAC 98]. However, we will analyse the performances of the maximum likelihood approach in the ill-posed setting under consideration (a small sample sizes, highly censored lifetimes, unknown failure origins). Without anticipating the numerical experimental results, there is a concern that the maximum likelihood methodology will provide large variance estimations. Thus, Bayesian inference could be expected to be useful.

Bayesian inference is now described in three steps. First, the prior distributions are defined; secondly, Gibbs sampling is designed, and finally an alternative adaptable importance sampling procedure is described.

11.4.1. *Choosing the prior distributions*

Four prior distributions will be chosen for the four model parameters: the scale parameters $(\eta_1; \eta_2)$ and the shape parameters $(\beta_1; \beta_2)$ from the two Weibull distributions. Since the two components of the competing risk model are assumed to be independent, the prior distributions can be chosen independently for each component. For the Weibull distribution there is no conjugate prior distribution for (β, η). However, for a fixed β, the gamma distribution is a conjugate prior for parameter η. Thus, it is natural to choose a prior on (β, η)

$$\pi(\beta, \eta) = \pi(\beta)\pi(\eta / \beta)$$

where $\pi(\eta/\beta)$ is a $G(a, b)$. The prior distribution of the shape parameter β is then chosen. Typically, this parameter is bound between two values β_g and β_d. A natural choice for $\pi(\beta)$ is then a beta distribution with parameters p and q for the interval $[\beta_g, \beta_d]$. The beta distribution is a versatile distribution and, by varying p and q, we can get quite different distributions (see [BAC 98], p.52 for illustrations). This prior distribution is denoted as $B(p, q, \beta_g, \beta_d)$.

11.4.2. *From prior information to prior hyperparameters*

The hyperparameters of the prior distributions a, b, p, q, β_g are β_d are fixed. They have to take into account the expert opinions and the uncertainty concerning those opinions in order to lead to a relevant statistical analysis. Since the mean of a Gamma distribution $G(a, b)$ is a/b and its variance is a/b^2, we consider an equation $a/b = m$ to enter the expert opinion on the mean value m of the component. Since the variance is a/b^2, b has to be chosen small enough to ensure that the expert opinion does not bring more information than the observed data. In the present case, the mean lifetime given by the expert was 125 for both components. This leads to choosing the same prior gamma $G(12.5, 0.1)$ for both components.

The differences between the components concerned, as is often the case, are the shape parameters of the Weibull distributions. The expert does not know if the first component will have teething problems and thinks that its ageing effect is moderate. This leads to choosing a prior distribution of $B(1.5, 2, 0.5, 3)$. Taking $\beta_g = 0.5$ means that a youth default occurrence remains possible and $\beta_d = 3$ corresponds to a somewhat standard value for ageing. The second component is known to have no more teething problems, but the expert has not indicated a precise opinion about ageing. This leads to the prior choice $B(1.2, 1.5,1, 5)$. Choosing $\beta_g = 1$ is a direct consequence of the absence of a youth default and $\beta_d = 5$ is a large value for the shape parameter of a Weilbull distribution in an industrial context. Finally, the resulting prior distributions appear to be moderately informative. It is expected that degenerate estimates for the scale parameters will be avoided, as can occur with maximum likelihood estimates from poorly informative and highly censored data. In fact, choosing two different priors for the shape parameters could help to contrast the two components.

11.4.3. *Gibbs sampling*

Starting from an initial guess $\theta^0 = (\beta_1^0, \eta_1^0, \beta_2^0, \eta_2^0) = (1.5, 100, 2.5, 120)$ the Gibbs sampler can be described as follows, with the exponent q denoting the iteration index:

– generate the failure labels e_i^q according to their conditional distribution, knowing the current value θ^q of the parameters. This conditional distribution is a Bernoulli distribution with parameter, for $i = 1, \ldots, r$

$$p(e_i^q = 1) = \frac{\beta_1^q / \eta_1^q (t_i / \eta_1^q)^{\beta_1^q - 1}}{\beta_1^q / \eta_1^q (t_i / \eta_1^q)^{\beta_1^q - 1} + \beta_2^q / \eta_2^q (t_i / \eta_2^q)^{\beta_2^q - 1}}.$$

– generate the shape parameters β_1^{q+1} and β_2^{q+1} according to their conditional posterior distributions, knowing the current values of the scale parameters and the failure labels e_i^q. Since the prior beta is not conjugate, these conditional posterior distributions are not standard and this step involves the use of the Metropolis–Hastings algorithm. An instrumental distribution is possibly the product of the maximum likelihood function and of the prior distribution of the parameter to be simulated;

– generate the scale parameters η_1^{q+1} and η_2^{q+1} according to their conditional posterior distributions, knowing the current values of the shape parameters and the failure labels e_i^q. Since, for fixed shape parameters, the prior distribution on the scale parameter is conjugate, this step becomes a simulation of a gamma distribution whose hyperparameters are dependent on the data, the simulated labels and the shape parameters values, and obviously of the chosen prior hyperparameters a and b.

Thus, Gibbs sampling takes advantage of the missing data structure of the competing risk model by simulating the missing labels of the components causing the failures with the model parameters, at each iteration. This occurrence of Gibbs sampling requires a Metropolis–Hastings step and choosing the instrumental distribution for that step can involve difficulty. In such cases, a random walk for this instrumental distribution from the previous values could be recommended since it leads to quicker convergence than the product of the likelihood and the prior.

11.4.4. *Adaptive Importance Sampling (AIS)*

The importance sampling we now present also takes advantage of the model's missing data. The procedure is detailed in [CEL 06]. It is an adaptive procedure mimicking Gibbs sampling by simulating the unknown labels and the parameters according to their conditional posterior distributions. The procedure is as follows, in which $e = (e_i, i = 1, \ldots, n)$:

Initialization: Choice of $(\theta_0^{(1)}, \ldots, \theta_0^{(M)})$ through M independent random samples taken from the prior distribution of θ.

$$\text{Step } q \ (q=1,...,Q): \quad p(e_i^q = 1) = \frac{\beta_1^q / \eta_1^q (t_i / \eta_1^q)^{\beta_1^q - 1}}{\beta_1^q \eta_1^q (t_i \eta_1^q)^{\beta_1^q - 1} + \beta_2^q \eta_2^q (t_i \eta_2^q)^{\beta_2^q - 1}}$$

(a) for $i = 1, \ldots, M$, we simulate $e_q^{(i)}$ and $\theta_q^{(i)}$ according to their conditional distributions. In practice, this stage consists of repeating the simulations performed with the Gibbs sampler M times. We then compute the weights, for $i = 1, \ldots, M$:

$$p_q^{(i)} = \frac{g(t, e_q^{(i)} / \theta_q^{(i)}) \pi(\theta_q^{(i)})}{k(e_q^{(i)} / t, \theta_{q-1}^{(i)}) \pi(\theta_q^{(i)} / t, e_q^{(i)})}$$

where $t = (t_1, \ldots, t_r, c, \ldots, c)$, $g(t,e)$ is the completed likelihood knowing the model parameters and $k(e/t, \theta)$ is the conditional density of the labels, knowing t and θ. These weights are then normalized, such that:

$$w_q^{(i)} = p_q^{(i)} / \sum_{j=1}^{M} p_q^{(j)}$$

(b) Simulation of $\theta_q^{(i)}$ according to the distribution defined by $w_q^{(i)}$s.

In practice M has to be large (several hundred). Here, we have taken $M = 500$. By contrast, the number of iterations, Q, for this adaptive procedure can be small. We have chosen $Q = 10$ here[1].

We now take the estimates produced by the EM algorithm, its stochastic version (SEM) in a maximum likelihood setting, then estimates obtained with the Gibbs sampling (GIBBS) and the adaptive importance sampling (AIS) described above, with the prior as detailed in section 11.4.2. The EM estimates are $\beta_1 = 1.59$, $\eta_1 = 109.4$, $\beta_2 = 1.64$ and $\eta_2 = 78.5$, and they are not very satisfying. Typically, these estimates highlight the tendency of EM to provide almost equal shape parameters for competing risk models.

From SEM, we get the following estimates (the standard errors due to the random simulations are given in brackets): $\beta_1 = 1.16$ (0.12), $\eta_1 = 94.6$ (6.15), $\beta_2 = 3.60$ (0.28) and $\eta_2 = 185.1$ (5.40). This illustrates the tendency of SEM to assign each point to the same component, implying an overestimation of neglected

---

1 It is possible that only a few weights w are non-zero, especially when the number of observed failures is small. To avoid this, the weights are normalized for each iteration [CMR 06].

components while the parameters of other components are well estimated. Finally, this case highlights the fact that maximum likelihood is jeopardized by small and poorly informative samples. In fact, EM and SEM give quite different estimates.

Table 11.1 displays the estimates of the posterior mean and posterior standard error for each component provided by the GIBBS and AIS methods. The columns m and σ give, respectively, the estimated posterior mean and standard error. Both methods provide analogous and satisfying estimates.

Reference	$\beta_1=1.5$		$\eta_1=100$		$\beta_2=2.5$		$\eta_2=120$	
	$m_{\beta 1}$	$\sigma_{\beta 1}$	$m_{\eta 1}$	$\sigma_{\eta 1}$	$m_{\beta 2}$	$\sigma_{\beta 2}$	$m_{\eta 2}$	$\sigma_{\eta 2}$
GIBBS	1.75	0.24	88.8	7.8	2.41	0.47	100.5	10.7
EPA	1.7	0.21	91.5	10.1	2.26	0.46	103.9	12.2
EM	1.59		109.4		1.64		178.5	
SEM	1.16		94.6		3.6		185.1	

Table 11.1. *Maximum likelihood estimates through EM and SEM algorithms, Bayesian posterior means and standard errors through GIBBS and AIS methods for the Weibull competing risk model*

This difference in performance between maximum likelihood and Bayesian estimates increases when the sample size and the number of observed failure times decrease. But, it is important to note that a good result from the Bayesian approach is highly dependent on reasonable choices for the prior densities (namely a prior with large variances and realistic mean values).

When no expert opinion is available, it is recommended that a *non-informative* prior distribution is used. In a small sample setting, this will lead to more stable and reliable estimates than the maximum likelihood estimates. But, clearly, more relevant estimates are expected when relevant prior information is well calibrated in a prior distribution. From this point of view, it is important to understand that any Bayesian inference has a price to be paid. MCMC algorithms or importance sampling procedures are expensive and difficult to properly control. Moreover, a good calibration of prior distributions requires careful sensibility analyses of the prior hyperparameters, after much trial and error.

11.5. Conclusion

In this chapter, our analysis of statistical inference from small data sets has restricted its attention to simple problems, and many questions have not been tackled. Regularization methods for multivariate decisional statistics, for example, have not been introduced. Is it possible to draw useful and trivial information from data when the number of variables is dramatically greater than the number of data? Answering this question is more and more important since it is now possible to get a lot of information about complex characters in genetics, image analysis, data from the web, etc. Regularization methods are, for example, used in the optimization of Kernel methods such as *Support Vector Machines* or regression models such as *ridge* regression, *Lasso* or *Elastic Net* methods, etc. This domain of statistical research receives a lot of interest today, and soon quite efficient regularization tools will be routinely available. Interested readers will find a clear and precise presentation of prominent regularization methods in [HAS 09].

The chapter has also shown the great interest in Bayesian inference when dealing with small samples. It is worth indicating a few books devoted to Bayesian inference. We have already cited the excellent [ROB 06] many times, which is essentially a French translation of the book in English [ROB 01]. Other less complete but valuable, and somewhat more affordable, books on Bayesian inference can also be recommended, such as [MAR 07], [PAR 07] or [PRO 09]. The next chapter of this book presents readers with an example of an application of Bayesian inference concerning warping prediction in a nuclear power plant's containment components. Finally, [ROB 04] contains a complete presentation of Monte Carlo methods in a Bayesian context, whilst [NTZ 09] has a description of MCMC method implementation in the Winbugs software and a concise self-contained entry on commented R programs.

11.6. Bibliography

[BAC 98] BACHA M., CELEUX G., IDEE E., LANNOY A., VASSEUR D. *Estimation de modèles de durées de vie fortement censurés,* Eyrolles, Paris, France, 1998.

[CEL 06] CELEUX G., MARIN J.-M., ROBERT C.P., "Iterated importance sampling in missing data problems", *Computational Statistics & Data Analysis*, 50, p. 3386-3404, 2006.

[DEM 77] DEMPSTER A., LAIRD N., RUBIN D. "Maximum likelihood from incomplete data via the EM algorithm", *Journal of the Royal Statistical Society*, B, 39, 1-38, 1977.

[ELA 06] EL ADLOUNI S., FAVRE A., BOBEE B., "Comparison of methodologies to assess the convergence of Markov chain Monte-Carlo methods", *Comput. Stat. Data Anal.* vol. 50(10), p. 2685-2701, 2006.

[HAS 70] HASTINGS W., "Monte-Carlo sampling methods using Markov chains and their application", *Biometrika*, vol. 57(1), p. 97-109, 1970.

[HAS 09] HASTIE T., TIBSHIRANI R., FRIEDMAN J., *The Elements of Statistical Learning* (2nd edition), Springer, New York, USA, 2009.

[MAR 07] MARIN J.-M., ROBERT C.P., *Bayesian Core: a Practical Approach to Computational Bayesian Statistics*, Springer, New York, USA, 2007.

[NTZ 09] NTZOUFRAS I., *Bayesian Modeling using Winbugs*, John Wiley & Sons, USA, 2009.

[PAR 07] PARENT E. BERNIER J., *Le raisonnement bayésien : modélisation et inférence*, Springer, Paris, France, 2006.

[PRO 09] PROCACCIA H., *Introduction à l'analyse probabiliste des risques industriels*, Tec et Doc, Paris, France, 2009.

[ROB 01] ROBERT C.P., *The Bayesian Choice: From Decision-Theoretic Motivations to Computational Implementation* (2nd edition), Springer, New York, USA, 2001.

[ROB 04] ROBERT C.P., CASELLA G., *Monte-Carlo Statistical Methods* (2nd edition), Springer, New York, USA, 2004.

[ROB 06] ROBERT C.P., *Le choix bayésien : principes et pratique*, Springer, Paris, France, 2006.

[ROB 09] ROBERT C.P., CASELLA G., *Introducing Monte-Carlo Methods in R* (2nd edition), Springer, New York, USA, 2009.

Chapter 12

Bayesian Updating Techniques in Structural Reliability

12.1. Introduction

Computer simulation models such as finite element models are nowadays commonly used in various industrial fields in order to optimize the design of complex mechanical systems as well as civil engineering structures. In the latter case, the structures under consideration (e.g. cable-stayed bridges, dams, tunnels, etc.) are often one of a kind. Thus, they are usually monitored during their construction and after, so that experimental measurement data (displacement, strains, etc.) are collected all along their lifetime.

These measurements are traditionally used to detect a possible unexpected behavior of the system (e.g. a temporal drift of some indicator). In this case, data is processed using classical statistical methods *without* any physical modeling of the structure. However, this data could be used together with a computational model elaborated at the design stage in order to update the model predictions. Classical approaches in this field are purely deterministic: the analyst tries to select the set of model parameters that best fit the available data by minimizing the discrepancy between measurements and model prediction (e.g. using least-square minimization), without taking into account possible sources of error or uncertainty such as measurement error, model uncertainty, etc.

In this chapter, section 12.1 describes a probabilistic framework that allows us to combine computational model (e.g. a finite element model), a prior knowledge on its

Chapter written by Bruno SUDRET.

input parameters and experimental database. This framework makes use of Bayesian statistics, which is rigorously presented Chapter 13. Section 12.2 presents a retained probabilistic model for the interaction between measurements and model predictions. Section 12.3 recalls how to compute the probability of failure of a structure and how to update it using additional measurement data. Section 12.4 describes how to "invert" the reliability problem and compute quantiles of the mechanical response instead. Section 12.5 presents the Markov Chain Monte Carlo (MCMC) simulation method – which has already been introduced in the previous chapter – and how it may be used for Bayesian updating problems. Finally, section 12.6 describes an application example related to the durability of concrete containment vessels in French nuclear power plants.

12.2. Problem statement: link between measurements and model prediction

Let us consider a mathematical model $\mathcal{M}(x,t)$ that represents the temporal evolution of the response $y(t) \in \mathbb{R}^N$ of a mechanical system as a function of a vector of input parameters $x \in \mathbb{R}^M$. These basic parameters are supposed to be uncertain or not well known. They are modeled by a random vector $X = \{X_1,...,X_M\}^\mathsf{T}$ whose joint Probability Density Function (PDF) is prescribed. This PDF may be selected at the design stage from available data (see Chapters 4 and 5) or from expert judgment. In this context, the model response at time instant t is a random vector denoted by $Y(t) = \mathcal{M}(X,t)$. The collection of random vectors $\{Y(t), t \in [0,T]\}$ is a random process (see Chapter 10 for a rigorous definition).

Let $\tilde{y}(t)$ be the "true" value of the system response at time instant t, i.e. the value that would be measured by a perfect, infinitely accurate device (with no measurement error). This value is usually different from the observed value $y_{obs}(t)$ which was obtained by the measurement device at hand.

If a perfect model of the system behavior was available, there would exist a vector of basic parameters denoted by $\tilde{x}$ such that $\tilde{y}(t) = M(\tilde{x},t)$. However models are always simplified representations of the real world and contain unavoidable approximations. Thus a so-called measurement/model error term is introduced in order to characterize the discrepancy between the model output and the corresponding observation. Considering various time instants $t^{(q)}, q = 1, \cdots, Q$, this assumption reads:

$$y_{obs}^{(q)} = \mathcal{M}\left(\tilde{x}, t^{(q)}\right) + e^{(q)} \qquad [12.1]$$

In the latter equation, the observed value $y_{obs}(t)$ and the (implicit or explicit) definition of the model $\mathcal{M}$ are known, whereas $\tilde{x}$ and $e^{(q)}$ are unknown. If one assumes that the error $e^{(q)}$ is a realization of a random vector $E^{(q)}$ that characterizes the measurement/model error (usually a Gaussian vector of zero mean value and covariance matrix C) the above equation means that $y_{obs}^{(q)}$ is a realization of a random vector $Y_{obs}^{(q)}$ whose *conditional distribution* reads:

$$Y_{obs}^{(q)}|X=x \sim \mathcal{N}(M(x,t^{(q)}),C) \tag{[12.2]}$$

where $\mathcal{N}(\mu,C)$ denotes a multinormal distribution with mean value μ and covariance matrix C.

In practice, depending on the problem under consideration, the error term $E^{(q)}$ may represent either the measurement uncertainty, the model error, or both. These two quantities are usually independent so that this error may be broken down again as the sum of two terms. The total covariance may be split as $C = C_{mes} + C_{mod}$, where C_{mes} represents the covariance matrix of the sole measurement error.

12.3. Computing and updating the failure probability

12.3.1. *Structural reliability – problem statement*

Structural reliability analysis aims to compute the probability of failure of a mechanical system whose parameters are uncertain and modeled within a probabilistic framework. Reliability methods that lead to computing a probability of failure with respect to a scenario are well documented in the books by Ditlevsen & Madsen [DIT 96] and Lemaire [LEM 09].

Let X denote the vector of input random variables describing the problem (which usually includes the input parameters of some mechanical model $\mathcal{M}$), and let us denote its support by $\mathcal{D}_X \subset \mathbb{R}^M$. A failure criterion may be mathematically cast as a limit state function, $x \in \mathcal{D}_X \mapsto g(x)$ such that $D_f = \{x : g(x) \leq 0\}$ is the failure domain and $D_s = \{x : g(x) > 0\}$ is the safe domain. The boundary between both domains is called the limit state surface, ∂D. The failure probability is then defined by:

$$P_f = \mathbb{P}(g(X) \leq 0) = \int_{D_f} f_X(x)\,dx \tag{[12.3]}$$

where f_X is the joint PDF of X.

As the integration domain D_f is implicitly defined from the sign of the limit state function g and the latter is usually not analytical, a direct evaluation of the integral in equation [12.3] is rarely possible. It can be numerically estimated using Monte Carlo simulation (MCS): N_{sim} realizations of the input random vector X are drawn according to its joint PDF f_X, and for each sample the g-function is computed. The probability of failure is estimated by the ratio N_f / N_{sim} where N_f is the number of samples (among N) that have lead to failure (i.e. a negative value of g).

This method, which is rather easy to implement, may be unaffordably costly in practice. Indeed, suppose that a probability of failure of the order of magnitude 10^{-k} is to be estimated with a relative accuracy of 5%: a number of $N_{sim} \approx 4.10^{k+2}$ simulations is then required. As failure probabilities usually range from 10^{-2} to 10^{-6} it is clear that MCS will not be directly applicable for industrial problems, for which a single run of the model $\mathcal{M}$ and the associated performance g may require hours of CPU. In order to bypass this difficulty, alternative approximate methods have been introduced such as the First Order Reliability Method (FORM).

FORM allows us to approximate the failure probability by recasting the integral in equation [12.3] in the standard normal space, i.e. a space in which all random variables ξ are normal with zero mean value and unit standard deviation. To this aim an isoprobabilistic transform $T : X \rightarrow \xi(X)$ is used.

If the basic random variables gathered in X are independent with respect to the marginal cumulative distribution function (CDF) $F_{X_i}(x_i)$, this transform reads: $\xi_i = \Phi^{-1}\left(F_{X_i}(x_i)\right)$, where Φ is the standard normal CDF. In the general case, the Nataf or Rosenblatt transforms may be used; see ([LEM 09], Chapter 4) for details. After mapping the basic variables X into standard normal variables ξ, equation [12.3] can be rewritten as:

$$P_f = \int_{\{\xi \,:\, G(\xi) \equiv g(T^{-1}(\xi)) \leq 0\}} \varphi_M(\xi)\,d\xi_1 ...d\xi_M \tag{12.4}$$

where $G(\xi) = g\left(T^{-1}(\xi)\right)$ is the limit state function in the standard normal space and φ_M is the multinormal (M-dimensional) PDF defined by

$\varphi_M(\xi) = (2\pi)^{-\frac{M}{2}} \exp\left[-\frac{1}{2}(\xi_1^2 + ... + \xi_M^2)\right]$. This PDF is maximal for $\xi = 0$ and decreases exponentially with $\|\xi^2\|$. Thus the points that contribute most to the integral in equation [12.4] are those points of the failure domain that are close to the origin of the space.

The next step of FORM consists of determining the so-called *design point* ξ^*, i.e. the point of the failure domain D_f that is closest to the origin. This point is the solution of the following optimization problem:

$$\xi^* = \underset{\xi \in \mathbb{R}^M}{\text{Arg min}} \left\{\frac{1}{2}\|\xi\|^2 \; / \; G(\xi) \leq 0\right\} \tag{[12.5]}$$

Dedicated constrained optimization algorithms may be used to solve it. The minimal (algebraic) distance from the limit state surface ∂D to the origin is called the Hasofer–Lind reliability index: $\beta = \text{sign}(G(0))\|\xi^*\|$. Once ξ^* has been computed, the limit state surface ∂D is linearized around this point and replaced by a tangent hyperplane. The failure domain is then substituted by the half space defined by this hyperplane. The approximation of the integral in [12.4] by integrating over the half space leads to the FORM approximation $P_f \approx P_{f,FORM} = \Phi(-\beta)$.

The equation of the linearized limit state (i.e. the hyperplane) may be cast as $\tilde{G}(\xi) = \beta - \alpha^\top \cdot \xi$. In this expression, the unit vector (which is orthogonal to the hyperplane) contains the cosines of the angles defining the direction of the design point. The square cosines α_i^2 are called *importance factors* since they allow one to break down the variance of the (approximate) performance $\tilde{G}(\xi)$ into contributions of each variable ξ_i and, by extension, to quantify the impact of each basic variable X_i onto the reliability.

The First Order Reliability Method allows the analyst to get an approximation of P_f at a reasonable computational cost (usually, from a few tens to a few hundreds of evaluations of g). The approximation is all the better since the reliability index β is large. Moreover, the approach yields importance factors. Sensitivity measures that quantify how the probability of failure changes when some assumption of the basic random variables is changed are also interesting quantitative indicators for the designer.

12.3.2. *Updating failure probability*

The failure probability, as defined in the previous section, is usually computed at the design stage. For an already existing system for which additional information is available (e.g. measurements of response quantities in time), it is possible to update the probability of failure by accounting for this data.

Let us consider a set of observations[1] $\mathcal{Y} = \left\{ y_{obs}^{(1)}, ..., y_{obs}^{(Q)} \right\}^{\mathrm{T}}$ collected at time instants $t^{(q)}$, $q = 1, ..., Q$ along the lifetime of the structure. Confronting this data with model simulation results leads to the introduction of so-called *measurement events* $\left\{ H_q = 0 \right\}$ [DIT 96] using the following notation:

$$H_q = \mathcal{M}\left(\boldsymbol{X}, t^{(q)} \right) - y_{obs}^q + E^{(q)} \tag{12.6}$$

In this equation $E^{(q)}$ denotes a Gaussian random variable that characterizes the measurement/model error. The updated failure probability $P_f^{upd}(t)$ is now defined as the following conditional probability:

$$P_f^{upd}(t) = \mathbb{P}\left(g\left(\boldsymbol{X}, t \right) \leq 0 \mid H_1 = 0 \cap ... \cap H_Q = 0 \right) \tag{12.7}$$

When recasting the measurement events as the limit, $\lim_{\theta \to 0} \left\{ -\theta < H_q \leq 0 \right\}$, we get:

$$P_f^{upd}(t) = \lim_{\theta \to 0} \frac{\mathbb{P}\left(\left\{ g\left(\boldsymbol{X}, t \right) \leq 0 \right\} \cap \left\{ -\theta < H_1 \leq 0 \right\} \cap ... \cap \left\{ -\theta < H_Q \leq 0 \right\} \right)}{\mathbb{P}\left(\bigcap_{q=1}^{Q} \left\{ -\theta < H_q \leq 0 \right\} \right)} \tag{12.8}$$

In the above equation both the numerator and denominator are failure probabilities of parallel systems (intersections of events) that may be estimated by an extension of the FOR method to systems ([LEM 09], Chapter 9). After some algebra, equation [12.8] reduces to [MAD 87]:

$$P_f^{upd}(t) = \Phi\left(-\beta^{upd}(t) \right) \text{ with } \beta^{upd}(t) = \frac{\beta_0(t) - z(t)^{\mathrm{T}} \cdot \boldsymbol{R} \cdot \boldsymbol{\beta}}{\sqrt{1 - \left(z(t)^{\mathrm{T}} \cdot \boldsymbol{R} \cdot z(t) \right)^2}} \tag{12.9}$$

---

1 From this point on, the response quantity under consideration and the associated measurements are taken to be *scalar* quantities, for the sake of simplicity.

In this equation $\beta_0(t)$ is the *initial* reliability index associated with the event $\{g(X,t) \leq 0\}$ and $\boldsymbol{\beta} = \{\beta_1, ..., \beta_Q\}^{\mathrm{T}}$ gathers the reliability indices related to the events $\{H_q \leq 0\}$. Moreover, $z(t) = \{z_1(t), ..., z_Q(t)\}$ is the vector of correlations between the linearized margins $\{H_q = 0\}$ and $\{g(X,t) = 0\}$ whose components are $z_j(t) = \boldsymbol{\alpha}_0(t) \cdot \boldsymbol{\alpha}_j$. Finally $\boldsymbol{R}$ is the correlation matrix of the linearized measurement margins, whose generic entry reads $R_{kl} = \boldsymbol{\alpha}_k \cdot \boldsymbol{\alpha}_l$. Thus the updated failure probability may be computed only from a set of FORM analyses.

12.4. Updating a confidence interval on response quantities

12.4.1. *Quantiles as the solution of an inverse reliability problem*

Suppose the random response of a mechanical model $Y(t) = \mathcal{M}(X,t)$ is of interest. Its variability may be fruitfully grasped through the computation of a confidence interval on the prediction, which means computing quantiles of $Y(t)$. For instance, a 95%-confidence interval (i.e. a range such that the probability of $Y(t)$ being in this range is 95%) is obtained by computing the 2.5% and 97.5% quantiles of $Y(t)$. As a consequence, the computation of α-quantiles $y_\alpha(t)$ defined by:

$$\mathbb{P}\big(Y(t) \leq y_\alpha(t)\big) = \alpha \; ; \; \alpha \in \,]0,1[\tag{12.10}$$

is of interest. By introducing the mechanical model $\mathcal{M}$ in the previous equation, we obtain $y_\alpha(t)$ as the solution to the following:

$$\mathbb{P}\big(\mathcal{M}(X,t) - y_\alpha(t) \leq 0\big) = \alpha \tag{12.11}$$

Equation [12.11] may be considered for each time instant t as an *inverse reliability problem* [DER 94], in which the value of a parameter (here, y_α) is looked for so that a given "failure probability" is attained (here, α) for a given limit state function (in this case, $\mathcal{G}(X,t \, ; \, y_\alpha) \equiv \mathcal{M}(X,t) - y_\alpha$). In order to solve this problem efficiently an extension of FORM has been proposed in [DER 94]. Within the FORM approximation the problem is recast as:

$$\text{find } y_\alpha : \; P_{f,FORM}\big(g(X,t \, ; \, y_\alpha) \leq 0\big) = \Phi(-\beta_c) \tag{12.12}$$

where $\beta_c = -\Phi^{-1}(\alpha)$ is the target reliability index associated with the α-quantile of interest. The algorithm used for computing quantiles is presented in detail in [PER 07], [PER 08].

12.4.2. *Updating quantiles of the response quantity*

The "inverse FORM" approach may be elaborated one step further in order to compute "updated quantiles", which are defined as quantiles computed conditionally to observations. When combining [12.9] and [12.12] the "updated" version of the latter reads:

$$\text{find } y_\alpha : \quad P_{f,FORM}\left(g(X,t\,;y_\alpha) \le 0\,|\,H_1 = 0 \cap ... \cap H_Q = 0\right) = \Phi(-\beta_c) \qquad [12.13]$$

where the measurement events are defined in equation [12.6]. The "updated inverse FORM" algorithm as originally proposed in [SUD 06] couples the inverse FORM algorithm with equation [12.9] by modifying in each iteration the target reliability index $\beta_c^{(k+1)}$ which is equal at iteration $k+1$ to:

$$\beta_c^{(k+1)} = -\Phi^{-1}(\alpha)\sqrt{1-\left(z^{(k)}(t)^{\mathrm{T}} \cdot R \cdot z^{(k)}(t)\right)^2} + z^{(k)}(t)^{\mathrm{T}} \cdot R \cdot \beta^{(k)} \qquad [12.14]$$

In the above equation, matrix R does not change from one iteration to the other, in contrast to vectors z and β. Note that the convergence of the algorithm is not proven, although numerous application examples have shown the efficiency of the method.

12.4.3. *Conclusion*

The method proposed in the previous section allows us to update the failure probability of a structure or, indirectly, to update the confidence intervals of the prediction of a mechanical model by using measurement data gathered all along the lifetime of the structure.

This approach enables the reconciliation of the prior model predictions $Y(t) = \mathcal{M}(X,t)$ and the observed data $\mathcal{Y} = \left\{y_{obs}^{(1)},...,y_{obs}^{(Q)}\right\}^{\mathrm{T}}$ in order to better estimate the probability of failure of the real structure ("as built") under consideration. However it does not bring any additional information to the basic variables X. An alternative approach based on Markov Chain Monte Carlo simulation is presented in the next section for this purpose.

12.5. Bayesian updating of the model basic variables

12.5.1. *A reminder of Bayesian statistics*

Bayesian statistical methods [OHA 04], [ROB 92] are usually used in order to combine prior information on parameters of a random vector with data, i.e. realizations, of that random vector. Let us denote by $\mathcal{X} = \left\{ x^{(1)},..., x^{(Q)} \right\}$ a set of observations that will be modeled by a PDF $f_X(x;\theta)$, where θ is the vector of hyperparameters of size n_θ. Bayesian statistics assumes that some prior information on θ exists that may be modeled by a prior distribution $p_\Theta(\theta)$ of support $D_\Theta \subset R^{n_\theta}$. Bayes' theorem, in its continuous setting, combines both sources of information in order to yield a *posterior distribution* $f_\Theta(\theta)$:

$$f_\Theta(\theta) = \frac{1}{c} p_\Theta(\theta) L(\theta ; \mathcal{X}) \qquad [12.15]$$

In this equation L is the likelihood of the observations which is defined in case of independent observations by:

$$L(\theta ; \mathcal{X}) = \prod_{q=1}^{Q} f_X\left(x^{(q)} ; \theta \right) \qquad [12.16]$$

and c is a normalizing constant defined by $c = \int_{D_\Theta} p_\Theta(\theta) L(\theta;\mathcal{X}) d\theta$. From [12.15] and [12.16] we can further obtain the predictive distribution of X, namely:

$$f_X^p(x) = \int_{D_\Theta} f_X(x,\theta) f_\Theta(\theta) d\theta \qquad [12.17]$$

More directly the point posterior distribution of X reads $\hat{f}_X(x) = f_X(x,\hat{\theta})$, where $\hat{\theta}$ is a characteristic value of the posterior distribution $f_\Theta(\theta)$, e.g. the mean or median value.

12.5.2. *Bayesian updating of the model basic variables*

As observed from equation [12.2], each measurement data may be modeled by a random variable whose conditional distribution with respect to the vector of input variables reads:

$$f_{Y_{obs}^{(q)}|X}\left(y\,;x,t^{(q)}\right)=\varphi_M\left(y-\mathcal{M}\left(x,t^{(q)}\right);C\right)$$

$$\equiv(2\pi)^{-M/2}(\det C)^{-1/2}\exp\left[-\frac{1}{2}\left(y-\mathcal{M}\left(x,t^{(q)}\right)\right)^{\mathsf{T}}\cdot C^{-1}\cdot\left(y-\mathcal{M}\left(x,t^{(q)}\right)\right)\right] \qquad [12.18]$$

Let us denote by $p_X(x)$ the prior distribution of the input random vector X, i.e. the one used in reliability analysis before introducing measurement data. Using Bayes's theorem, we can evaluate the posterior distribution denoted by $f_X(x)$ through the likelihood of the measurement data gathered in y :

$$f_X(x)=\frac{1}{c}\,p_X(x)L(x\,;\mathcal{Y})=\frac{1}{c}\,p_X(x)\prod_{q=1}^{Q}\varphi_M\left(y_{obs}^{(q)}-\mathcal{M}\left(x,t^{(q)}\right);C\right) \qquad [12.19]$$

The normalizing constant c in the above equation ensures that $f_X(x)$ is a distribution (of integral 1). Its computation may be carried out using simulation methods (such as Monte Carlo simulation, Latin Hypercube Sampling, etc.) or numerical integration (e.g. the Gauss quadrature method). However, this is a rather complex computational task.

Another approach consists of sampling according to this posterior distribution by using a method that does *not* require the computation of the normalizing constant c. This is one feature of the so-called Markov Chain Monte Carlo simulation methods presented in the previous chapter.

Various algorithms such as the Gibbs sampler or the Metropolis–Hastings algorithm [HAS 70] are available; see a review in [NTZ 09]. The Metropolis–Hastings algorithm is an acceptance/rejection algorithm that works as follows. Suppose a random vector X of prescribed PDF $f_X(x)$ is to be sampled, and suppose the PDF has a complex expression that may be evaluated for any value x up to a constant. A Markov chain is initiated (value $x^{(0)}$). At the current state of the chain $x^{(k)}$ at iteration k, the next point $x^{(k+1)}$ is evaluated as follows:

$$x^{(k+1)}=\begin{cases}\tilde{x}\sim q\left(x\,|\,x^{(k)}\right) & \text{with probability } \alpha\left(x^{(k)},\tilde{x}\right),\\ x^{(k)} & \text{otherwise}\end{cases} \qquad [12.20]$$

In this equation $q(x|x^{(k)})$ is the *transition* (or *proposal*) distribution that is selected by the analyst and $\alpha(x^{(k)},\tilde{x})$ is the *acceptance probability*. A common transition is obtained by generating the candidate $\tilde{x}$ by adding to each component a random disturbance to $x^{(k)}$ according to a prescribed (e.g. zero-mean Gaussian) distribution.

$$\tilde{x}_i = x_i^{(k)} + \zeta_i^{(k)} \quad ; \quad \zeta^{(k)} \sim \mathcal{N}\left(0, \sigma^2\right) \tag{12.21}$$

This is the so-called random walk algorithm. In this case the acceptance probability is reduced to:

$$\alpha(\mathrm{x}^{(k)}, \tilde{x}) = \min\left\{1, \frac{f_X(\tilde{x})}{f_X(\mathrm{x}^{(k)})}\right\} \tag{12.22}$$

In order to decide whether the candidate point $\tilde{x}$ is retained with the acceptance probability $\alpha(x^{(k)}, \tilde{x})$, a random number $u^{(k)}$ is uniformly sampled between 0 and 1. The candidate is accepted if $u^{(k)} < \alpha(x^{(k)}, \tilde{x})$ and rejected otherwise. Thus a sequence of points is simulated, which is proven to behave asymptotically as realizations of the random vector X. We must check if the Markov chain has attained its stationary state, i.e. that a sufficiently large number of points has been simulated. Various heuristic control methods have been proposed in the literature; see for instance a review in [ELA 06].

The Metropolis–Hastings algorithm may also be used in a cascade version in which the candidate point is first accepted or rejected with respect to the ratio of prior distributions, then with respect to the likelihood ratio. The algorithm proposed by Tarantola [TAR 05] for this purpose is now described.

[Initialization] $k = 0$: The Markov chain is initialized by $x^{(0)}$ which can be randomly selected or deterministic (i.e. the vector mean value).

While $k \leq N_{MCMC}$ (N_{MCMC} is the size of the MCMC sample set) **DO**

1.   Generate   a   random   increment $\zeta^{(k)} \sim N(0, \sigma^2)$ and   a   candidate $\tilde{x} = x^{(k)} + \zeta^{(k)}$.

2. Evaluate the prior acceptance probability: $\alpha_P(x^{(k)}, \tilde{x}) = \min\left\{1, \frac{f_X(\tilde{x})}{f_X(x^{(k)})}\right\}$

3. Randomly generate $u_P \sim \mathcal{U}[0,1]$. If $u_P < \alpha_P(x^{(k)}, \tilde{x})$ then $\tilde{x}$ is accepted (*Go to 4.*) otherwise it is rejected (*Go back to 1.*)

4.   Evaluate   the   likelihood   acceptance   probability

$$\alpha_L(x^{(k)}, \tilde{x}) = \min\left\{1, \frac{L(\tilde{x}; \mathcal{Y})}{L(x^{(k)}; \mathcal{Y})}\right\},$$ where the likelihood function L has been

defined in equation [12.19]. This step requires a run of the deterministic model $\mathcal{M}$.

5. Randomly generate $u_L \sim \mathcal{U}[0,1]$. If $u_L < \alpha_L(x^{(k)}, \tilde{x})$ then $\tilde{x}$ is accepted: $x^{(k+1)} \leftarrow \tilde{x}$ and $k \leftarrow k+1$. Otherwise $\tilde{x}$ is rejected.

Coming back to the initial problem of updating the predictions of a model by using observation data, the MCMC algorithm is applied in a cascade to the posterior distribution of the random vector X, as defined in equation [12.19]. The sample set of points that is obtained, say $\mathcal{X}' = \left\{x^{(1)}, \cdots, x^{(N_{MCMC})}\right\}$, is then used as input of a Monte Carlo simulation of the model $\mathcal{M}$. In practice the evaluations of $\mathcal{M}$ onto the sample set $\mathcal{X}'$ have already been carried out during the process of generating $\mathcal{X}'$. Computing the updated confidence intervals of the model prediction reduces to an estimation of the related empirical quantiles on the already available response sample set $\left\{\mathcal{M}(x^{(1)}), \cdots, \mathcal{M}(x^{(N_{MCMC})})\right\}$.

In conclusion, the Bayesian approach based on Markov Chain Monte Carlo simulation allows us to update the distribution of the input random vector by incorporating the observations made on the system response. From this updated (i.e. posterior) distribution, updated confidence intervals may be computed that compare with those obtained by the "updated inverse FORM" algorithm. Both approaches will now be benchmarked on an industrial example in the following section.

12.6. Updating the prediction of creep strains in containment vessels of nuclear power plants

12.6.1. Industrial problem statement

The containment vessel of a nuclear power plant contains the reactor pressure vessel and the components of the primary circuit, namely pumps, steam generators and pipes. The leak tightness of this vessel is guaranteed in case of a hypothetical accident such as a loss of coolant accident (LOCA) that could happen when a pipe is ruptured, thus generating a rapid pressure increase within the vessel while possibly releasing radioactive products from the primary circuit.

The containment vessels of French pressurized water reactors (PWRs) are made of one or two walls, each made of reinforced and pre-stressed concrete. The so-called concrete creep phenomenon, which corresponds to delayed strains in concrete due to ageing, leads to a decrease of the tension of the pre-stressing cables over time. In order to assess the safety of the containment vessel all along the lifetime of the plant in the context of hypothetical LOCA accidents, it is necessary to accurately predict the evolution over time of the delayed stresses and associated loss of cable pre-stress.

However, the creep phenomenon is very complex in nature. Its physical origins are not fully understood, especially when its kinetics over long-term time scales is concerned. In order to bypass the lack of detailed modeling, a detailed monitoring of the containment vessels has been installed. Thus measurements of the delayed strains in standard conditions are carried out on a regular basis. The Bayesian framework that has been presented in the previous sections is well adapted to exploit this experimental feedback together with physical models of creep.

12.6.2. *Deterministic models*

Let us consider a cylindrical portion of the containment vessel that is sufficiently far away from local geometrical details (reinforcements, material hatch, etc.) so that it is relevant to consider that the concrete stress tensor under cables pre-stress is bi-axial (the pre-stress cables are vertical and circumferential in this zone). The mechanical model used in the sequel for delayed stresses is defined in the French standard BAEL [BAE 99] although it takes into account specific modifications as investigated by Granger for containment walls [GRA 95].

Accordingly the total strain tensor ε can be broken down into the elastic, creep and shrinkage components:

$$\varepsilon(t,t_d,t_l) = \varepsilon^{el}(t) + \varepsilon^{as}(t,t_d) + \varepsilon^{ds}(t,t_d) + \varepsilon^{bc}(t,t_l) + \varepsilon^{dc}(t,t_d,t_l) \qquad [12.23]$$

where:

– t is the time spent starting from the concrete casting, t_d denotes the time when drying starts and t_l denotes the time of loading, i.e. cable tensioning in the present case;

– $\varepsilon^{el}(t)$ is the elastic strain;

– $\varepsilon^{as}(t,t_d)$ is the autogenous shrinkage, corresponding to the shrinkage of concrete when insulated from humidity changes;

– $\varepsilon^{ds}(t,t_d)$ is the drying shrinkage;

– $\varepsilon^{bc}(t,t_l)$ is the basic creep corresponding to the creep of concrete when insulated from humidity changes;

– $\varepsilon^{dc}(t,t_d,t_l)$ is the drying creep.

The following models are used for each component. The elastic strains are related to the stress tensor σ by Hooke's law:

$$\varepsilon^{el} = \frac{1+v^{el}}{E_i}\sigma - \frac{v^{el}}{E_i}(\operatorname{tr}\sigma)\,I \qquad [12.24]$$

where E_i is the elastic Young's modulus (measured at $t=t_l$) and v^{el} is the Poisson's ratio. The autogenous and drying shrinkage are modeled (with a time unit of one day) by:

$$\varepsilon^{as}(t,t_d) = \varepsilon^{as}_{\infty}\frac{t-t_d}{50+t-t_d}\,I \qquad \varepsilon^{ds}(t,t_d) = \varepsilon^{ds}_{\infty}\frac{100-RH}{50}\frac{t-t_d}{45R^2_m/4+t-t_d}\,I$$

$$[12.25]$$

In these equations $\varepsilon^{as}_{\infty}$ is the asymptotic autogeneous shrinkage and $\varepsilon^{ds}_{\infty}$ the asymptotic drying shrinkage, RH is the relative humidity in %, R_m is the drying radius (half of the containment wall thickness, in cm) and I is the unit strain tensor, meaning that these strains are isotropic. The basic creep is modeled by:

$$\varepsilon^{bc}(t,t_1) = 3500\left(\frac{1+v^c}{E_i}\sigma - \frac{v^c}{E_i}(tr\sigma)I\right)\left(\frac{2{,}04}{0{,}1+(t_i-t_d)^{0{,}2}}\right)\left(\frac{\sqrt{t-t_1}}{22{,}4+\sqrt{t-t_1}}\right)$$

$$[12.26]$$

where v^c is the creep Poisson's ratio. The drying creep is modeled by:

$$\varepsilon^{dc}(t,t_d,t_l) = 3200\frac{\operatorname{tr}\sigma/2}{E_i}\left(\varepsilon^{ds}(t,t_d)-\varepsilon^{ds}(t_l,t_d)\right)I \qquad [12.27]$$

In a pre-stressed concrete containment vessel, the stress tensor may be regarded as bi-axial in the current zone, i.e. having a vertical component $\sigma^0_{zz} = 9.3$ MPa and an orthoradial component $\sigma^0_{\theta\theta} = 13.3$ MPa. The drying radius, which is equal to half of the wall thickness, is 0.6 m. The cable tensioning is supposed to occur two years after the casting of concrete ($t_l - t_d = 2$ years).

Due to the presence of reinforcing bars and pre-stressed cables, the above equations for creep and shrinkage (initially obtained for unreinforced concrete) are corrected by a multiplying factor $\lambda = 0.82$ obtained from the design code and experimental results [GRA 95]. The other parameters are modeled by independent random variables, whose parameters are gathered in Table 12.1.

Parameter	Notation	Distribution	Mean value	Coefficent of variation
Concrete Young's modulus	E_i	LogNormal	33,700 MPa	7.4 %
Poisson's ratio	ν^{el}	Truncated normal [0; 0.5]	0.2	50 %
Creep Poisson's ratio	ν^c	Truncated normal [0; 0.5]	0.2	50 %
Relative humidity	RH	Truncated normal [0; 100%]	40%	20 %
Max. autogenous shrinkage strain	ε_∞^{as}	LogNormal	90×10^{-6}	10 %
Max. drying shrinkage strain	ε_∞^{ds}	LogNormal	526×10^{-6}	10 %

Table 12.1. *Concrete creep model – probabilistic description of the parameters*

A fictitious containment vessel is considered for which it is supposed that experimental measurements of the axial strain ε_{zz} are available. Measurements are supposed to have been carried out approximately every 150 days from 1,500 and 2,500 days after the concrete structure was loaded. They are reported in Table 12.2.

The measurement/model error is supposed to be normally distributed with zero mean value and standard deviation 15.10^{-6}. The various errors at different time instants are assumed to be independent.

Measurement	Date (days)	Value (10^{-6})
#1	1152	497
#2	1303	523
#3	1451	590
#4	1601	652
#5	1750	685
#6	1900	756
#7	2054	777
#8	2201	822
#9	2153	858
#10	2501	925

Table 12.2. *Concrete creep model – fictitious strain measures*

12.6.3. *Prior and posterior estimations of the delayed strains*

All the simulation results have been obtained using the probabilistic model reported in Table 12.1. The prior 95% confidence interval on the vertical strain ε_{zz} is computed by the "inverse FORM" approach. Results are gathered in Figure 12.1, in which the measurement values from Table 12.2 have also been plotted. These results have been validated by Brute Force Monte Carlo simulation [PER 08].

It can be observed that using the prior estimation of the parameters' distribution (Table 12.1) leads to a large underestimation of the vertical delayed strains of c.40%. This can be explained by the fact that the values of the prior model parameters have been taken from a building code (BAEL) and thus are not well adapted to the specific concrete used for containment vessels.

The "updated inverse FORM" approach is then applied using the measurement values in Table 12.2. Results are plotted in Figure 12.2. It can be observed that the posterior 95% confidence interval now covers the experimental data and that it is much smaller than the prior interval. The Bayesian framework has allowed us to reconcile the experimental data with the model and to reduce the uncertainty in the prediction of the long-term behavior of the structure. It has been shown in [SUD 06] that the posterior result is not very sensitive to the number of data used for the updating process since the time variation of creep is rather slow and smooth.

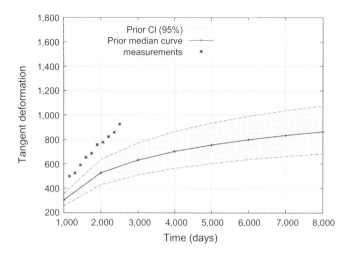

Figure 12.1. *Prior predictions of the vertical total strain ε_{zz} and fictitious experimental results*

The MCMC approach for updating the distributions of the input parameters has also been applied. The results are plotted in Figure 12.3 and corroborate those obtained by the "inverse FORM" approach, the maximal discrepancy between the updated quantiles obtained by each approach being less than 4%. The obtained updated confidence interval is slightly tighter than that obtained by inverse FORM (Figure 12.2) and four times smaller than the prior interval.

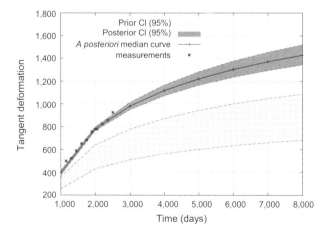

Figure 12.2. *Prior/posterior predictions of the vertical total strain ε_{zz} obtained by the "updated inverse FORM" approach and experimental results*

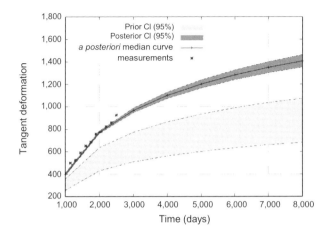

Figure 12.3. *Prior/posterior predictions of the vertical total strain ε_{zz} obtained by MCMC and experimental results*

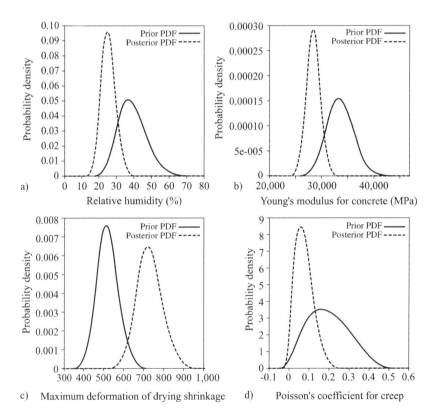

Figure 12.4. *Prior and posterior distributions of selected random variables*

As indicated in section 12.5, the MCMC method yields the posterior distribution of the various input random variables of the problem. Some of these distributions are plotted in Figure 12.4. We can observe that all the posterior distributions are less scattered than the corresponding priors: adding information within the computational model has reduced the uncertainty.

12.7. Conclusion

Structural reliability methods are usually used when designing a complex structure so as to guarantee that the failure probability associated with various criteria is sufficiently low. For exceptional civil engineering structures such as cable-stayed bridges or nuclear concrete containment vessels, monitoring is usually established from construction of the system. Thus a large amount of data is collected all along the lifetime of the system. In this chapter it has been shown that this data may be used in order to refine the long-term evolution of the structure.

The Bayesian updating techniques presented in this chapter allow the analyst to address this question efficiently. The "inverse FORM" approach only updates the quantiles of the model response, without yielding any additional information on the model input variables. By contrast, the Markov Chain Monte Carlo simulation allows the distributions of the input variables to be updated from a prior to a posterior estimate. The latter posterior distributions may be re-propagated through the mechanical model in order to obtain updated quantiles.

The same methods have been used successfully to predict the crack propagations in steel structures [SUD 07] and the delayed strains of a concrete containment vessel by using a detailed (finite element) model for creep and shrinkage [BER 11]. In the latter case, the computational cost of a single run of the model is rather large. Thus a surrogate model of the finite element model was built first, namely a polynomial chaos expansion (see Chapter 8 for details). This surrogate, which is essentially a polynomial function of the input variables, may then be used straightforwardly within the MCMC algorithm.

In conclusion, it should not be forgotten that the Bayesian framework, although an elegant approach for integrating experimental feedback into computational models, cannot completely replace proper physical modeling: in particular, a physical model $\mathcal{M}$ should at least describe the general trend of the time evolution of structural behavior in order to obtain relevant results.

12.8. Acknowledgments

The application of Bayesian updating methods to the monitoring of civil engineering structures is one of the research topic of the MIRADOR project (*Modélisation Interactive et Recalage par l'Auscultation pour le Développement d'Ouvrages Robustes*) that has been supported by the French National Research Agency (*Agence Nationale de la Recherche)* under Grant ANR-06-RGCU-008. This support is gratefully acknowledged.

12.9. Bibliography

[BAE 99] BAEL, Règles BAEL 91, modifiées 99 (Règles techniques de conception et de calcul des ouvrages et constructions en béton armé suivant la méthode des états-limites), Eyrolles, Paris, France, 2000.

[BER 11] BERVEILLER M., LE PAPE Y., SUDRET B., PERRIN F., "Updating the long-term creep strains in concrete containment vessels by using MCMC simulation and polynomial chaos expansions", *Struc. Infrastruc. Eng.*, vol. 7, 2011.

[DER 94] DER KIUREGHIAN A., ZHANG Y., LI C., "Inverse Reliability Problem", *J. Eng. Mech.*, vol. 120, p. 1154-1159, 1994.

[DIT 96] DITLEVSEN O., MADSEN H., *Structural Reliability Methods*, John Wiley & Sons, Chichester, 1996.

[ElA 06] EL ADLOUNI S., FAVRE A., BOBEE B., "Comparison of methodologies to assess the convergence of Markov chain Monte Carlo methods", *Comput. Stat. Data Anal.*, vol. 50(10), p. 2685-2701, 2006.

[GRA 95] GRANGER L., Comportement différé du béton dans les enceintes de centrales nucléaires, PhD thesis, Ecole Nationale des Ponts et Chaussées, Paris, France, 1995.

[HAS 70] HASTINGS W., "Monte Carlo sampling methods using Markov chains and their application", *Biometrika*, vol. 57(1), p. 97-109, 1970.

[LEM 09] LEMAIRE M., *Structural Reliability*, ISTE Ltd, London and John Wiley & Sons, New York, USA, 2009.

[MAD 87] MADSEN H., "Model updating in reliability theory", LIND N. (ed.), *Proc. 5th Int. Conf. on Applications of Stat. and Prob. In Civil Engineering (ICASP5)*, Vancouver, Canada, vol. 1, p. 564-577, 1987.

[NTZ 09] NTZOUFRAS I. *Bayesian Modeling Using Winbugs*, John Wiley & Sons, New York, USA, 2009.

[OHA 04] O'HAGAN A., FOSTER J., *Kendall's Advanced Theory of Statistics, Vol. 2B Bayesian Inference*, Arnold, London, 2004.

[PER 07] PERRIN F., SUDRET B., PENDOLA M., DE ROCQUIGNY E., "Comparison of Markov chain Monte Carlo simulation and a FORM-based approach for Bayesian updating of mechanical models", *Proc. 10th Int. Conf. on Appl. of Stat. and Prob. In Civil Engineering (ICASP10)*, Tokyo, Japan, 2007.

[PER 08] PERRIN F., Intégration des données expérimentales dans les modèles probabilistes de prévision de la durée de vie des structures, PhD thesis, Blaise Pascal University, Clermont-Ferrand, France, 2008.

[ROB 92] ROBERT C., *L'analyse statistique bayésienne*, Economica, Paris, France, 1992.

[SUD 06] SUDRET B., PERRIN F., BERVEILLER M., PENDOLA M., "Bayesian updating of the long-term creep deformations in concrete containment vessels", *Proc. 3rd Int. ASRANet Colloquium*, Glasgow, UK, 2006.

[SUD 07] SUDRET B., Uncertainty propagation and sensitivity analysis in mechanical models – Contributions to structural reliability and stochastic spectral methods, Habilitation à diriger des recherches, PhD thesis, Blaise Pascal University, Clermont-Ferrand, France, 2007.

[TAR 05] TARANTOLA A., *Inverse Problem Theory and Methods for Model Parameter Estimation*, Society for Industrial and Applied Mathematics (SIAM), 2005.

Reliability-based Maintenance Optimization

Introduction to Part 5

To cope with structural degradation, maintenance operations aim to preserve an acceptable level of reliability over all the desired service life. Preventive maintenance allows us to considerably reduce failure probability but leads to additional costs which are, in many cases, due to benefit losses resulting from system unavailability. The objective of a maintenance policy is therefore to search for the best compromise between the conflicting requirements of cost and reliability. An optimal maintenance plan has to ensure the best reliability/availability at the lowest possible cost.

This part of the book aims to provide an understanding and basic illustration of the application of reliability-based maintenance optimization. Chapter 13 introduces maintenance concepts by describing various types of maintenance to be applied, with the corresponding maintenance models to be used, in order to improve structural reliability. Chapter 14 provides a detailed look at maintenance cost models and describes criteria for choosing the appropriate policy, where an illustrative example is provided for tubular pipe under corrosion. Finally, Chapter 15 presents two industrial applications and discusses the limits of this type of study in the context of multiple criteria, focusing on studies of the maintenance of a highway concession and of a cooling tower in a nuclear power plant, to illustrate the main concepts presented in Chapters 13 and 14.

Chapter 13

Maintenance Policies

13.1. Maintenance

Maintenance can be defined as the combination of operations allowing the system being considered to be maintained in a desired state of performance. The possible operations include inspections, repairs and replacements. This chapter defines various types of maintenance, as well as some definitions related to structural lifetime, T. Remember that the failure probability is denoted as P_f and reliability by $R = 1 - P_f$. The failure rate, given as $\lambda(t)$, and measured at time t, can be simply defined as the number of failed items per unit time. Readers can refer to a number of works [LAN 05], [MOR 01], [NAK 02] for more details concerning the concepts related to the evaluation and management of ageing industrial systems, in a wider scope than just for construction.

13.1.1. *Lifetime distribution*

Distribution of lifetime T, denoted f_T, can be estimated either by a physics approach or by statistical approaches. For most industrial systems, the specification of "mean lifetime" m_T does not present a useful indicator (Figure 13.1), as a very high failure probability leads to undesired consequences in terms of economic and life losses. By contrast, the specification of "guaranteed lifetime", corresponding to very low failure rate, implies the under use and underating of a system, and consequently leads to economic losses that could be avoided. We can therefore define the "technical–economic lifetime" as the service time behind which either the

Chapter written by Alaa CHATEAUNEUF, Franck SCHOEFS and Bruno CAPRA.

industrial risk is considered very high, or required additional investments cannot be refunded during the system's future life. An optimal compromise can be found on the basis of this technical–economic lifetime, corresponding to the minimum expected total cost. This solution can be obtained by optimal balancing of initial and additional investments, risk of losses and costs of monitoring and maintenance actions.

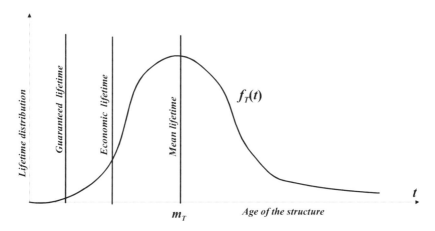

Figure 13.1. *Lifetime distribution*

13.1.2. *Maintenance cycle*

At the beginning of a service life, a structure is in a state of good operation until the occurrence of its first failure (Figure 13.2). In many cases, a system is partially or totally repairable (or replaceable) and can re-start a new life-cycle until the next failure. The consideration of failure in service is generally accompanied by some administrative and technical delay which should be taken into account before starting effective repair operations which themselves require a specific time interval. Knowing that uncertainties cannot be avoided, the duration corresponding to each state is intrinsically random and, consequently, we should not only consider the mean values but also the associated dispersions. Figure 13.2 illustrates the evolution of a system in service, and we can distinguish the following periods:

– Mean Time To (first) Failure (MTTF): corresponding to the expected time for the occurrence of the first failure after the beginning of the system's operation. For unrepairable systems, the MTTF corresponds to the expected lifetime;

– Mean Time Between Failures (MTBF): corresponding to the expected time between two successive failures of the same component/structure, including the down time during maintenance (i.e. this time is defined by the interval between the moments of the occurrence of successive failures). The MTBF is divided into an operation period, known as the Mean Up Time (MUT), and a non-operating period, known as Mean Down Time (MDT);

– Mean Up Time (MUT): corresponding to the expected operation time from the end of repair till the next failure;

– Mean Down Time (MDT): corresponding to the expected time during which the system is stopped, from the occurrence of failure till the effective re-start of operation after repair. MDT itself can be divided into two parts: Mean Waiting Time (MWT) and Mean Time To Repair (MTTR);

– Mean Waiting Time (MWT): corresponding to the lapse of time before discovering the failure, for administrative procedures and for waiting for ordered or transported spare parts;

– Mean Time To Repair (MTTR): corresponding to the time necessary to carry out the repair operation, after having the material and human resources, until the total recovery of the service.

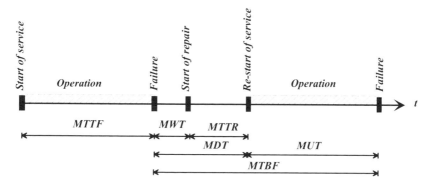

Figure 13.2. *Maintenance cycle for repairable systems*

13.1.3. *Maintenance planning*

The utility of a structure depends on its capability to ensure good operation on one hand, and the socio-economic conditions, on the other hand. The capability of ensuring good operation is generally sensitive to the use conditions, the environment (especially climatic and geotechnical), the quality applied (materials, methods of analysis, transport or construction, etc.) and the procedures of maintenance and inspection. The use conditions are often affected by climatic, geotechnical and socio-economic environment. The maintenance itself is affected by the environment and by the applied quality (Figure 13.3).

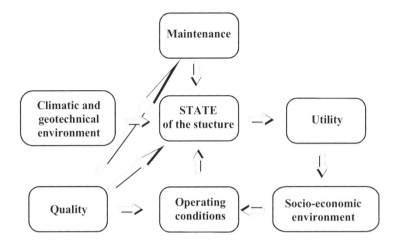

Figure 13.3. *Inter-dependence of the conditions defining a structure's utility*

Maintenance planning should be based on equilibrated combination of preventive and corrective operations, in coherence with inspection results. The maintenance policy provides the sequence of decisions at various stages of the structure's life. At the design stage, the maintenance parameters should be defined in an optimal way. At the operation stage, the designer and the user/owner should specify the inspections and actions in order to allow for better knowledge and management of the structure's lifetime and performance (Figure 13.4).

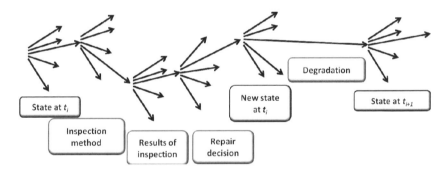

Figure 13.4. *Decision tree for inspections*

The inspection parameters are the time span between successive inspections on one hand, and the depth of those inspections on the other hand. The inspection depth depends on the methods to be applied and their quality (i.e. technique, operator qualifications, etc.). The inspection results should lead to enough information to allow a decision to be made, with more or less precision, about whether maintenance actions should or should not be performed. The main objective in this tree consists of minimizing the maintenance cost[1], by taking into account safety constraints, and respect of standards, codes and regulations, without forgetting the limits of the available budget.

Therefore, the objective of a maintenance policy is to maintain the structure in a state of "good operation". The aim is to keep "the observed reliability during operation", called "operational reliability", higher than the "target reliability", called "intrinsic reliability". This reliability-based maintenance policy consists of defining the tasks related to the following aspects:

– failure modes;

– failure causes;

– detection tools;

– maintenance tools;

– procedures, methods and systems provided to realize maintenance;

– quantities (in terms of the outcome of maintenance and inspections), as well as the corresponding decisions.

13.2. Types of maintenance

Various classifications of maintenance action appear in the literature. According to Figure 13.5, the maintenance can be either preventive or corrective. Preventive maintenance can be either systematic, according to a time schedule defined in advance, or conditional, continuous or on request, or scheduled as a function of the information collected during the service life. At the same time, corrective maintenance can be either delayed, if the system can operate without risk until the scheduled time, or performed urgently, when repair is necessary to ensure safety and good operation.

13.2.1. *Choice of the maintenance policy*

The choice of a maintenance policy depends on at least three criteria: (i) the impact of maintenance on the requirements of safety, availability and cost; (ii) the data characterizing the degradation; and (iii) the quality of performed inspections. When the impact of preventive maintenance is low, we prefer to wait until failure; corrective maintenance must be adopted for failed components (Figure 13.6).

---

1 The terminology of optimization under maintenance constraint (i.e. minimization of the expected total cost or maximization of expected utility) will be specified in Chapter 14.

However, if the failure cost is high, then preventive maintenance becomes necessary. If the mean time to failure can be estimated, verified or checked, preventive maintenance can be planned at regular time intervals (whether calendar time, real operation time, or any other specific time function).

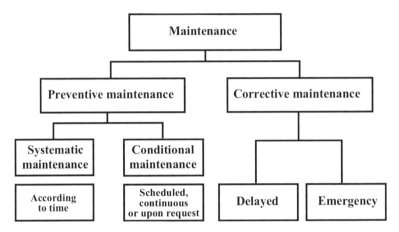

Figure 13.5. *Classification of various types of maintenance [CEN 01]*

If the degradation can be inspected or monitored, preventive maintenance should be based on the knowledge of the real state of the system. However, in some cases, the programming of maintenance is not possible, due to large uncertainties regarding failure rates. We may face this situation when the induced failure cost is very high although the state of the system cannot be inspected, or when the failures are recurrent with low costs. In this latter case, the system should be re-designed in order to ensure appropriate robustness.

In addition to safety obligations, which are often mandatory and dominant, preventive maintenance is justified by the economic consequences when the following two conditions are satisfied:

– the preventive operations, or inspections, planned at fixed dates, cost globally less than the inevitable random corrective operation that could sometimes be performed at critical moments;

– the failure rate of the system for which we intend to apply a program of preventive maintenance, increases due to degradation of the components of the system with an increase in financial consequences.

In continuous production sectors, such as energy production plants and petrochemical plants, economic dependence is a result of production losses when the system is down, during corrective or preventive maintenance. In the case of corrective downtime, additional costs have to be supported to take into account the direct and indirect consequences of failure (i.e. delay, de-damaging, urgent operation, etc.). During preventive or corrective operations, maintenance allows either the complete restoration of the components concerned or just the repair of the failure, with or without modifications of the ageing mechanisms. During corrective maintenance of a component, it is possible to take the opportunity of replacing other components of the same system preventively; in that case, we talk about "opportunistic maintenance".

Finally, maintenance actions remain possible right until the stage of high degradation, whether this involves regular inspections of a system to know its state and the quality of its previous maintenance, in order to decide about the correction to perform on the observed damage, or what to repair in order to bring the system back to a state close to its initial state, or, ultimately, the replacement of components when they reach a state of technical obsolescence.

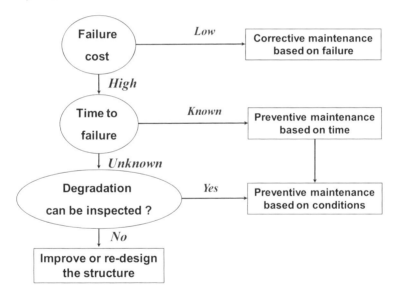

Figure 13.6. *Flowchart of the choice of maintenance policy*

These actions satisfy the economic requirements which should be taken into account in the development of the most recent theory of decision. The risk associated with each degradation and failure mode, has to be probabilistically evaluated. Possible actions like Risk-Based Inspection (RBI) [FAB 02a], Reliability-Based Maintenance (RBM), and Reliability-Centered Maintenance (RCM), repair or decommissioning have to be considered with their related uncertainties and possible consequences (i.e. economic losses, human fatalities, social impact, etc.), in order to determine the best possible decisions concerning the type of actions, including renovation and recertification of the inspected structure.

13.2.2. *Maintenance program*

A maintenance program consists of actions performed preventively (systematic or conditional) or correctively [ZWI 96]. Although systematic maintenance is performed at pre-defined times, conditional maintenance is performed when the acceptance criterion is reached, and corrective maintenance is performed urgently when failure occurs. Observations and inspection results lead to identification of the following states: acceptable operation, system to be monitored, or failed system. On the basis of the available information, (inspection results, experience feedback, expert opinion, etc.), the possible decisions concern the programming of alternatives (i.e. tasks of preventive and corrective maintenance) are made.

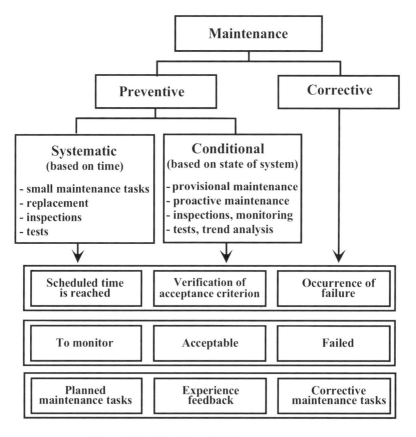

Figure 13.7. *Flowchart of maintenance program*

13.2.3. *Inspection program*

An inspection program starts by defining the system and the performance criteria (Figure 13.8); this step implies the collection of data concerning the system and its environment. A Qualitative Risk Analysis (QRA) is therefore performed to determine the failure modes, as well as their causes and consequences (using, for example, RAMS methods, as described in Part One). In this step, the logical representation of the system must be established, in order to define the required functions and the interactions between components and sub-systems. The next step consists of estimating the failure rates and consequences, in order to allow for ranking according to risk levels.

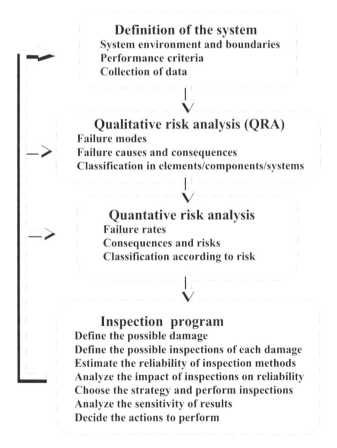

Figure 13.8. *Flowchart of an inspection plan. ("Defining the possible damage" implies the definition of degradation mechanisms; counting failures and degradations is usually sufficient, compared to evaluation of the failure rate)*

We can therefore distinguish between two cases:

– the case where defect measurement is sufficient to indicate the level of degradation and performance (direct evaluation of the limit state); and

– the case where measurement, using models, can help to evaluate the limit state (by measurement of material properties, degradation parameters, etc.); this case is most commonly observed in real systems.

An inspection program can be defined in terms of available tools and methods for each type of degradation. It is also important to take into account the positive impact (improvement of the state of knowledge; see "Model 1" below) or negative impact (system degradation) of inspections on the failure probability of the system. A comparison between different strategies allows us to make a judicious choice and to study the sensitivity of the results in terms of inspection parameters, leading to decisions regarding the maintenance plan and actions to be undertaken.

An inspection result is itself a product of modeling and operation, for a given context. We can therefore mention two families:

– Model 1: the inspection indicates the state of the system with negligible error.

– Model 2: *in situ* inspection is affected by non negligible errors, either at the detection level, or in the measured defect.

Besides updating the limit state function, Model 1 allows us to update the degradation model itself. In particular, we can use the collected information to update the probabilistic model of the measured parameter. This can be performed, for example, by means of Bayes theorem (Part Four, Chapters 10 to 12). Model 2 is more frequently met in industrial cases. It requires a probabilistic modeling of the inspection process itself (see Chapter 2). In fact, much research has been carried out to measure the possibilities of updating the probabilistic model in this case, though the problem is not considered in this book.

13.3. Maintenance models

Preventive and corrective actions can be classified in terms of their effect on the operational performances of a structural system:

– perfect maintenance: As Good As New (AGAN), where the system is brought back to its new state;

– minimal maintenance: As Bad as Old (ABAO), where the system is brought to its state just before failure;

– imperfect maintenance: Better than Old (BTO), where the performances are improved significantly;

– bad maintenance: Worse than Old (WTO), where performances deteriorate after maintenance.

It is clear that the classification should also take into account the physical aspects of the structure and the complexity of the system, in addition to the performed actions (repair, replacement of failed components, general or partial revisions, etc.). If the system is more or less complex, the maintenance is rather ABAO than AGAN.

It should be noted that maintenance models can also be established on the basis of the As Low As Reasonably Practicable (ALARP) concept.

13.3.1. Model of perfect maintenance: AGAN

A perfect maintenance is one which restores a system to its new state; we can say that the system becomes "as good as new" (AGAN). This model is particularly convenient for replacement of structures or members without any effect on the rest of the system (Figure 13.9).When perfect maintenance is performed at time t_i, the state of the system depends only on its history between t_i and t. Knowing that the system is perfectly repaired at each failure, its state can be described by "a renewal point process": the failure rate $\lambda(\cdot)$ is therefore a function of the time elapsed since the last failure/repair action:

$$\lambda(t \,|\, H_t) = h(t - t_i) \qquad\qquad [13.1]$$

where $h(\cdot)$ is the hazard function, H_t is the history of the structure, t indicates the time and t_i is the time at the last operation.

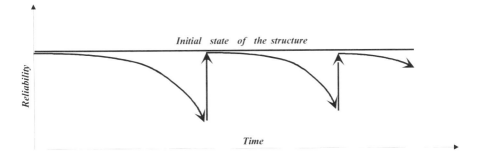

Figure 13.9. *"As good as new" maintenance model*

This scheme implies that the time intervals between successive failures $z_i = t_i - t_{i-1}$ are independent variables which are identically distributed. There is no trend for evolution of this variable with time. A special form of this process corresponds to a constant failure rate. In that case, the time intervals between failures are exponentially distributed. The assumption of perfect maintenance is not always verified by observations on real ageing systems; it is generally unproven to say that a repair operation makes a system new, unless we only look at it at the component or member scale.

13.3.2. *Model of minimal maintenance: ABAO*

Minimal repair is generally a corrective maintenance operation allowing a system to be restored to its state just before failure; we can say that the system becomes "as bad as old" (ABAO). When minimal maintenance is performed at t_i, the failure rate $\lambda(\cdot)$ after repair is identical to the one just before failure:

$$\lambda\left(t_i^+ \,\middle|\, H_{t_i^+} \right) = \lambda\left(t_i^- \,\middle|\, H_{t_i^-} \right)$$ [13.2]

where t_i^- and t_i^+ are respectively the times just before the *i*-th failure and just after the *i*-th repair.

In this type of maintenance, the reliability function is not affected by the events: failure–repair and the system have the property of "memory loss". If the system is subject to minimal maintenance at each failure without applying preventive maintenance, the failure rate remains a continuous function for the whole lifetime of the structure and does not depend on the history of the process, but only on the operating time:

$$\lambda(t \,|\, H_t) = \rho(t)$$ [13.3]

where $\rho(t)$ is the failure intensity, defined as the number of failures per unit time.

The assumption of minimal repair leads to a Poisson process (i.e. a counting process where the increments are independent and distributed according to Poisson law). The occurrence probability of k events in the interval $[t, t + \Delta t]$ can be written as:

$$P\left[N(t, t + \Delta t) = k \right] = \frac{\left[M(t, t + \Delta t) \right]^k}{k!} \exp\left(-M(t, t + \Delta t) \right) \quad for \quad k = 0, 1, 2, \ldots$$

where $N(t, t + \Delta t)$ is the number of failures in the interval $[t, t + \Delta t]$ and $M(t, t + \Delta t)$ is the mathematical expectation of this number during the same interval, obtained by: $M(t, t + \Delta t) = E[N(t, t + \Delta t)] = \int_t^{t + \Delta t} \rho(u)\, du$

Let Z be the Poisson process modeling the evolution of time intervals between failures. The random variables $Z_i = Z(t_i)$ can be neither identical nor independent. The conditional probability density function $f_{Z_i}(z \mid t_{i-1})$ of Z_i depends on the last failure time, noted as t_{i-1}:

$$f_{Z_i}(z \mid t_{i-1}) = \rho(t_{i-1} + z)\, exp\left[-\int_0^z \rho(t_{i-1} + u)\, du \right] \quad \text{for} \quad z \geq 0 \qquad [13.4]$$

Contrary to renewal processes, only the first interval $[0, t_1]$ is characterized by a failure rate equal to $\rho(t)$.

13.3.3. *Model of imperfect or bad maintenance: BTO/WTO*

In most cases, maintenance operations lead to an improvement of the state of the system, without making it new; we say that the system is "better than old" (BTO). Among the possible reasons for this situation, we can mention imperfections in repair methods, defects induced by the operation itself, the effect of human factors, the impact of the procedures on the rest of the system (e.g. assembly, disassembly), etc. In some cases, maintenance can be bad when the induced errors and defects are significant, if the maintenance staff is not qualified or has insufficient experience, or the working conditions are difficult or inappropriate; in such cases we say that the maintenance is "worse than old" (WTO). Another reason for BTO/WTO maintenance is the application of a "minimal repair time" policy instead of a "minimal repair" policy, in order to increase system availability. Several models have been developed to describe this type of maintenance.

13.3.3.1. *Arithmetic Reduction of Age (ARA) model*

The Arithmetic Reduction of Age (ARA) model consists of a reduction of the structural age with an amount proportional to the operation time elapsed since the last maintenance action [DOY 04]. In other words, each maintenance operation allows us to reduce the damage developed since the previous maintenance. The age reduction can be obtained by the expression:

$$\lambda(t \mid H_t) = h(t - \gamma t_{i-1}) \quad for \quad 0 \leq \gamma \leq 1 \quad and \quad t_{i-1} \leq t \leq t_i \qquad [13.5]$$

where γ is the efficiency parameter. In this case, the failure rate contains jumps at each maintenance operation. The special cases of minimal or perfect repairs are obtained when the efficiency parameter is equal to 0 or 1, respectively.

13.3.3.2. *Non Homogeneous Gamma Process (NHGP)*

The effect of imperfect or inefficient maintenance can be considered by assuming that the system is subject to jumps, also called "shocks", which happen according to a Non Homogeneous Poisson Process (NHPP) of intensity $\rho(t)$. The failure takes place, not at all the jumps, but only at the k-th jump. For $k > 1$, the component is in a better state after maintenance and, consequently, the process describes the case of imperfect maintenance. However, for $k < 1$, the state of the component is degraded after each maintenance operation (i.e. in the case of WTO maintenance). The random variable "numbers of jumps": $N_i = \int_{t_{i-1}}^{t_i} \rho(t)dt$ is independent and identically distributed according to Gamma law with identical scale and shape parameters, equal to k. The failure rate is given by:

$$\lambda(t|H_t) = \frac{\rho(t)\left(E\left[N(t_{i-1},t)\right]\right)^{k-1} \dfrac{exp\left(-E\left[N(t_{i-1},t)\right]\right)}{\Gamma(k)}}{\int_t^\infty \rho(z)\left(E\left[N(t_{i-1},z)\right]\right)^{k-1} \dfrac{exp\left(-E\left[N(t_{i-1},z)\right]\right)}{\Gamma(k)} dz} \qquad for \quad t > t_{i-1} \quad [13.6]$$

with $E[N(t_{i-1},t)] = \int_{t_{i-1}}^t \rho(z)dz$ the expectation of the number of jumps between the last failure and the time t. The process indicated above is called a Non Homogeneous Gamma Process (NHGP). For $k = 1$, this process is reduced to a Non Homogeneous Poisson Process (NHPP), corresponding to minimal repair; when $\rho(t) = c$, the process becomes a renewal process with times between failures distributed according to Gamma distribution, with scale parameter c and shape parameter k.

13.3.3.3. *Lawless-Thiagarajah model*

An interesting form of the failure rate which allows the inclusion of time dependence and the effect of observed failures can be written as:

$$\lambda(t|H_t) = \exp\left(\sum_{i=1}^n \theta_i \, g_i(t)\right) \qquad\qquad [13.7]$$

where θ_i are the model parameters and $g_i(t)$ are functions depending on the time and on the history of the process H_t. Particular forms of this equation can be written:

– $\lambda(t|H_t) = \exp(\alpha + \beta t + \gamma u(t))$ with $-\infty < \alpha, \beta, \gamma < \infty$

– $\lambda(t|H_t) = \gamma t^{\beta-1}(u(t))^{\delta-1}$ with $\gamma > 0,\ \beta + \delta > 0$

– $\lambda(t|H_t) = \delta \exp(\theta + \beta t)[u(t)]^{\delta-1}$ with $-\infty < \theta, \beta < \infty,\ \delta > 0$

where $u(t)$ is a time function. The two last forms of the failure rate are, respectively, associated with renewals of the Weibull and Weibull log-linear models.

13.3.3.4. *Arithmetic Reduction of Intensity (ARI) model*

Another way to take into account the beneficial effect of maintenance consists of reducing the failure rate, either by a constant quantity, or proportional to the failure rate before maintenance. The intensity reduction can be obtained by:

$$\lambda\left(t_i^+ \,\middle|\, H_{t_i^+}\right) = \delta\ \lambda\left(t_i^- \,\middle|\, H_{t_i^-}\right) \qquad with \quad \delta > 0 \qquad\qquad [13.8]$$

The parameter δ indicates the maintenance efficiency, while $\delta < 1$ corresponds to imperfect maintenance, $\delta > 1$ indicates worse than old maintenance, and $\delta = 1$ corresponds to minimal repair. At the ith failure, the failure rate is written as:

$$\lambda(t|H_t) = \delta^i\ \rho(t) \qquad with \quad t_i < t \le t_{i+1} \qquad\qquad [13.9]$$

where $\rho(t)$ is the failure intensity before the first failure.

13.3.4. **Complex maintenance policy**

A complex maintenance policy is made up of a combination of elementary maintenance actions, in order to minimize the cost or to increase availability. Various maintenance policies can be introduced, as summarized in Figure 13.10.

13.3.4.1. *Sequence of perfect and minimal repairs, without preventive maintenance*

Consider the following case: a structure is repaired when failure occurs, without preventive actions. The system is assumed to be subject to two types of failure: catastrophic with probability p and minor with probability $1-p$. The latter is corrected with minimal repair. A Non-homogeneous Poisson Process (NHPP) starts at each perfect repair. The distribution of times between successive perfect repairs is given as: $F_p(t) = 1 - [1 - F_1(t)]^p$ and the corresponding failure rate is: $\lambda_p(t) = p\lambda_1(t)$. The intensity is given by:

$$\lambda(t | H_t) = h(t - t_{i-1}) \quad \text{with} \quad t_{i-1} < t \le t_i \quad\quad [13.10]$$

This quantity is numerically equal to the failure rate of the first failure, with age given as $t - t_{i-1}$.

13.3.4.2. *Minimal repairs with perfect preventive maintenance*

In this scheme, preventive maintenance renews the state of a structure. In each time interval, a minimal repair is performed when failure occurs. In other words, at each maintenance cycle, a piecewise Non-Homogeneous Poisson Process (NHPP) can be defined. The variations of operating and environmental conditions between different cycles can be considered by introducing a covariance between the cycles:

$$h_{i,j}(t) = h_0(t) \exp\left(\sum_j c_j t_{i,j} \right) \quad\quad [13.11]$$

where $h_{i,j}(t)$ is the failure rate of the "time at the jth failure" during the "i^{th} cycle of maintenance", $t_{i,j}$ are the corresponding times and c_j are regression coefficients.

13.3.4.3. *Imperfect repairs with perfect preventive maintenance*

This scheme consists of performing repairs of type BTO when failure occurs, combined with component renewal during preventive maintenance of type AGAN. This policy represents the case of repairs in difficult conditions, the case of incident events, and complete replacement at planned times as part of preventive maintenance.

Type of maintenance	Policy 1	Policy 2	Policy 3	Policy 4
Corrective Maintenance	Perfect or minimal ⬇	Minimal ⬇	Imperfect ⬇	Minimal ⬇
Preventive Maintenance	Nothing	Perfect	Imperfect	Imperfect

Figure 13.10. *Examples of maintenance policies*

13.3.4.4. *Minimal repair with imperfect preventive maintenance*

This policy is probably the closest one to real structures. A minimal repair can always be expected in case of failure, while preventive maintenance cannot often be perfect. This policy aims at minimizing the down time due to failures and consequently maximizing system availability.

13.4. Conclusion

This chapter has introduced the vocabulary related to maintenance of structural components and infrastructures. Corrective or preventive maintenance can be divided into many types of actions, to be selected in an optimal way. The rich literature available should be consulted to provide simple analytical models. These models should be completed to include degradation mechanisms which are defined by complex phenomena with large uncertainties. A solution can therefore be obtained by applying numerical methods and optimization algorithms. The probabilistic approach allows us to take account of uncertainties related to degradation processes and to inspection and maintenance operations.

A probabilistic model of failures and degradations can be a decision-making tool for engineers and decision-makers, allowing the selection of an appropriate maintenance strategy. This choice can be a delicate one, for example, choice between valid techniques for which there is field experience and new techniques for which other advantages are interesting (in terms of performance, economy, etc.).

Finally, throughout the optimization process, we have to be aware of the comparison between the efficiency of various maintenance operations, and the consequences of heavy and delayed investments.

13.5. Bibliography

[CEN 01] CEN, EN 13306, Maintenance Terminology, 2001.

[DOY 04] DOYEN L., GAUDOIN O., "Classes of imperfect repair models based on reduction of failure intensity or virtual age", *Reliability Engineering & System Safety,* vol. 84(1), p. 45-56, 2004.

[ELL 97] ELLINGWOOD B.R., MORI Y., "Reliability-based service life assessment of concrete structures in nuclear power plants: optimum inspection and repair", *Nucl. Eng. Des.*, vol. 175, no. 3. p. 247-258, 1997.

[FAB 02a] FABER M.H. "RBI: An Introduction", *Structural Engineering International*, 3/2002, p. 187-194, 2002.

[LAN 05] LANNOY A., PROCACCIA H., *Evaluation et maîtrise du vieillissement industriel*, TEC & DOC, Lavoisier, Paris, France, 2005.

[MOR 01] MORI Y., NONAKA M., "LRFD for assessment of deteriorating existing structures", *Structural Safety*, 23, p. 297-313, 2001.

[NAK 02] NAKANISHI S., NAKAVASU H., "Reliability design of structural system with cost effectiveness during life cycle", *Comp. and Industrial Engrg*, 42, p. 447-456, 2002.

[ZWI 96] ZWINGELSTEIN G., *La maintenance basée sur la fiabilité*, Hermès Science, Paris, France, 1996.

Chapter 14

Maintenance Cost Models

14.1. Preventive maintenance

Preventive maintenance allows us to ensure an acceptable level of reliability during the structural lifecycle. A conditional maintenance policy is based on periodic inspections of degradation, which eventually trigger alarms related to repairs and replacements. In practice, however, quantitative knowledge of the state of degradation and operation conditions presents many uncertainties, which leads to a difficult decision-making process. Therefore, reliability-based maintenance becomes mandatory for decision-making.

The total maintenance cost can be written in the form:

$$C_M = C_F + C_{PM} + C_{INS} + C_{REP} \qquad [14.1]$$

where C_M is the expected total maintenance cost, C_F is the expected failure cost (including operation losses, production losses, and the direct and indirect damages due to failure), C_{PM} is the expected preventive maintenance cost, C_{INS} is the expected inspection cost, and C_{REP} is the expected repair cost. These costs are affected by uncertainties related to the state of degradation of the structure, the results of inspections and to repair/replacement methods. Moreover, these parameters may vary in terms of socio-economic environment, such as the discount rate, inflation and the fluctuations of market prices.

Chapter written by Alaa CHATEAUNEUF and Franck SCHOEFS.

The failure cost is related to direct damage (human lives, economic losses, loss of benefits, environment degradation, etc.) and to indirect damage (procedure fees, commercial impact, market losses, expert works, long term effects, etc.). Depending on the industry concerned, some costs may increase in an exponential way in terms of the failure rate. For example, the public relations/marketing impact, and therefore market losses, can jump considerably when the number of failed products becomes significant (which is the case in mass production, the automotive industry, aeronautics, etc.) or because the perception of risks makes them unacceptable (which is the case for nuclear power plants, dams, railways, etc.). In energy production industries (e.g. power plants, petro-chemical plants, etc.) the main losses are due to benefit losses when production is stopped. Moreover, during the last decade, the public has become more sensitive to aspects related to the environment; failures inducing pollution are severely punished by justice, politics and public opinion.

Figure 14.1 illustrates the costs of preventive and corrective maintenance, in terms of the level of planning. A low level of planning leads to larger number of emergency repairs and consequently to a larger total cost. Conversely, a high level of planned maintenance costs more and leads to losses due to over-maintenance. The optimization of the total maintenance cost allows us to find the best equilibrium in terms of the risks to be considered.

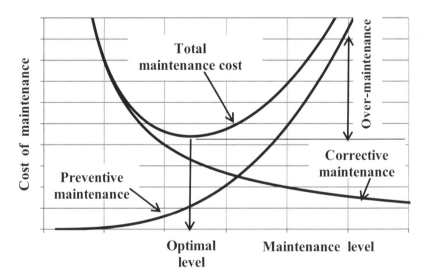

Figure 14.1. *Illustration of maintenance costs*

For a maintenance cycle $[0, \tau]$ under the assumption of an infinite horizon[1], the expectation of the maintenance cost per unit time takes the form:

$$C_M(\tau) = \frac{C^c P_f(\tau) + C^P R(\tau)}{\tau} \qquad [14.2]$$

where τ is the maintenance interval, C^c is the corrective maintenance cost, C^P is the preventive maintenance cost[2] and $P_f(\tau)$ is the accumulated failure probability at time τ; reliability is given by the survival probability $R(\tau) = 1 - P_f(\tau)$. The choice of maintenance strategy depends on the ratio between preventive and corrective costs; when $C^P < C^c$, preventive maintenance becomes useful, otherwise, maintenance should only be performed when the component fails.

14.2. Maintenance based on time

Under this policy, preventive maintenance is performed periodically at predefined times $k\tau$. When a component fails during an interval $[(k-1)\tau; k\tau]$, corrective maintenance is performed at the failure time. The advantage of this policy lies in its simplicity of application for the management of industrial systems, as preventive maintenance is previously planned and there is no need to monitor the ageing of components. Three versions of the model (I, II, III) can be considered:

– Model I: the failed component is instantaneously replaced when failure occurs;

– Model II: the failed component remains unrepaired until the next preventive maintenance;

– Model III: the failed component is subject to minimal repair until the next preventive maintenance.

----

1 The hypothesis of infinite horizon admits that the maintenance cycles are repetitive and identical at each renewal of the system. By contrast, the hypothesis of finite horizon admits that each maintenance cycle is different from a stochastic or economic point of view.

2 Unlike the notation C_{PM} and C_F which indicate the mathematical expectation of preventive maintenance and failure costs, respectively. The notation C^P and C^c indicate the deterministic costs of preventive and corrective maintenance (including the cost of failure), respectively. In other words, C^P and C^c are paid only when an event occurs.

14.2.1. *Model I*

In this model, the failed component is replaced by a new one during the maintenance interval, and all the components are systematically replaced at time intervals τ. According to renewal theory, the cost per unit time is obtained by:

$$C_M(\tau) = \frac{C^c E[N(\tau)] + C^p}{\tau} \qquad [14.3]$$

with $E[N(\tau)] = \sum_{i=1}^{\infty} P_f^{(i)}(\tau)$ being the expectation of the number of failures in the interval $[0;\tau]$ and $P_f^{(i)}(\tau)$ indicating the probability of having i failures in the interval (*i.e.* $P_f^{(i)}(\tau) = P[N(\tau) = i]$).

Under the assumption of only one failure of the same component during the interval, the above equation takes the form:

$$C_M(\tau) = \frac{C^c P_f(\tau) + C^p}{\tau} \qquad [14.4]$$

with C^c being the corrective maintenance cost and C^p the preventive maintenance cost.

14.2.2. *Model II*

In Model I, the component failure is immediately detected when failure occurs. In the absence of monitoring devices, it can be assumed that failure is detected only at the planned maintenance times $k\tau$. In Model II, the failed component remains unusable or unoperating after failure, until its detection. The expectation of the duration between failure and detection times is given by: $\int_0^\tau P_f(t)\,dt$. Therefore, the maintenance cost per unit time is written as:

$$C_M(\tau) = \frac{C^{ct} \int_0^\tau P_f(t)\,dt + C^p}{\tau} \qquad [14.5]$$

with C^{ct} being the corrective cost per unit time.

14.2.3. *Model III*

In Model III, it is assumed that the component undergoes minimal repair when failure occurs. The process of the number of failures $N(t)$ is not perturbed by the failure–repair couple of events. A Non-Homogeneous Poisson Process (NHPP) represents the behavior, where the expectation of the failure rate is given by:

$$\Lambda(t) = \int_0^t h(u)\,du \qquad [14.6]$$

with $h(t) = f(t)/R(t)$, called "the hazard function" and $\Lambda(t)$ "the cumulated hazard function". The total cost per unit time is therefore:

$$C_M(\tau) = \frac{C^{cm}\,\Lambda(\tau) + C^p}{\tau} \qquad [14.7]$$

where C^{cm} indicates the corrective cost of minimal repair.

14.3. Maintenance based on age

In this model, only the components which have survived till the planned time of preventive maintenance are replaced by new components, otherwise replacement is performed at failure. The advantage of this model is particularly significant when the only considered actions are replacement by a new component (i.e. repair is not considered to be alternative). The total cost per unit time is therefore:

$$C_M(\tau) = \frac{C^c P_f(\tau) + C^p R(\tau)}{\int_0^\tau R(t)\,dt} \qquad [14.8]$$

When the investment is damped for a long duration, it is necessary to update the cost by a factor r, called the "discount rate", taking into account the interest rate, inflation and other economic parameters. The present value of a cost is obtained by multiplying by the discount function $1/(1+r)^t$, which can be replaced by a continuous function $\exp(-rt)$. Under the assumption of infinite horizon, the updated expected cost is given by:

$$C_M(T) = \frac{C^c \int_0^T e^{-rt} f(t)\, dt + C^p e^{-rT} R(T)}{r \int_0^T e^{-rt} R(t)\, dt} \qquad [14.9]$$

14.4. Inspection models

14.4.1. *Impact of inspection on costs*

An optimal maintenance policy minimizes the total cost, including inspection and failure costs. A small time span between inspections leads to high costs of inspection operations, while a large time interval does not allow for timely detection of failures, therefore increasing the possibility of failure.

The first modeling consists of performing inspections at specific times, where each inspection is considered as instantaneous and perfect. The policy considers the inspection cost c_i and the failure cost c_f. The expected total inspection cost is written:

$$C_{IC} = \sum_{k=0}^{\infty} \int_{t_k}^{t_{k+1}} \left[c_i(k+1) + c_f(t_{k+1} - t) \right] dP_f(t) \qquad [14.10]$$

The solution of this leads to a recursive equation of the time intervals between inspections:

$$t_{k+1} - t_k = \frac{P_f(t_k) - P_f(t_{k-1})}{f(t_k)} - \frac{c_i}{c_f} \qquad [14.11]$$

with the first interval defined as: $c_i = c_f \int_0^{t_1} P_f(t)\, dt$

To simplify the solution method, some approximate formulations of the total cost have been proposed in the literature.

If we let $n(t)$ be the approximate number of inspections per unit time, the expected cost until the detection of the failure can be approximated by:

$$C_{IC}(n(t)) = c_i \int_0^{\infty} n(t)\, R(t)\, dt + c_f \int_0^{\infty} \frac{1}{2n(t)}\, dP_f(t) \qquad [14.12]$$

which can be minimized to give:

$$n(t) = \sqrt{\frac{c_f}{2c_i}} h(t) \qquad with \qquad h(t) = f(t) / R(t) \qquad\qquad [14.13]$$

The inspection times thus satisfy the equation: $k = \int_0^{t_k} n(t)\, dt$ where k is an integer.

14.4.2. *The case of imperfect inspections*

In situ inspection of structures is performed in conditions that are far from the ideal conditions found in a laboratory. When the operator has an important influence on the inspection result (precision and disposition of the material, visual reading, etc.), the working conditions directly affect the measurements. External factors such as fog, extreme temperatures, difficult working positions or internal factors such as fatigue and concentration level can be mentioned here as examples. In such cases, we talk about imperfect inspections.

We can use a probabilistic format to define the corresponding quantities of Probability of Detection (PoD) of a defect, and Probability of False Alarm (PFA). The calibration of these probabilities can be performed, either on the basis of statistical analysis, or by signal analysis. It should be noted that, in the case of PFA = 0, a Bayesian updating can be performed on the inspection results in order to modify the distribution of defects after inspection.

The technical performance of Non-Destructive Testing (NDT) devices and the chain of decision processes to achieve the information required are generally observed from two objectives regarding (i) the presence of defects (capacity of detection) and (ii) the measurement of the defect (capacity of measuring the physical or geometrical properties, such as the length and the depth of a crack). We can easily understand that a measurement (e.g. in the case of a lock or immersed piles of a wharf) realized at a number of meters depth is subject to large uncertainties, related to the following events:

– the diver gives s signal to the ground operator (beginning time of measurement t_0);

– the diver can handle the NDT device in operation more or less easily (depending on complexity of inspected joints, agitation due to waves and marine currents);

– the diver's vision is strongly reduced;

– the quality of the decision is based on the quality of cleaning of the surface to be inspected, especially of bio-dirtiness;

– diver fatigue and respiration difficulties come to increase the above difficulties.

Details on the available techniques, and their respective advantages and disadvantages, for the example of offshore platforms can be found in [ROU 01].

14.4.2.1. *Basic concepts and the Non-Destructive Testing (NDT) approach*

The application of the concept of Probability of Detection (PoD) was first raised in the 1980s [MAD 87], and then became more popular in the middle of the 1990s, especially in the planning of inspections according to the Risk Based Inspection (RBI) approach [FAB 02a], [FAB 02b], [MOA 97]. [MOA 98], [MOA 99].

Let us assume that a crack has been detected and we are considering the measurement of uncertainties. Let a_d be the detection threshold, that is, the size under which no crack can be detected. If unknown the distribution of defect d is called signal and measured defect "d hat" signal plus noise. The noise is the mathematical notion that allows us to model the errors of measurement, interpretation, see section 14.4.2. The probability of detection of a measured random defect $\hat{d}$ is therefore defined by:

$$\text{PoD}(a) = P\left[\hat{d} \geq a_d\right]$$
[14.14]

This definition is practical as long as the defect a can be described by a random variable. However, in the operational framework of inspection of real structures, defects are generally classified by groups and we prefer a Bayesian definition [ROU 03]:

$$\text{PoD}(X) = P\left[d(X) = 1 | X = 1\right]$$
[14.15]

where X is the event of "defect existence" and $d(.)$ the event "decision". The realization "$X = 1$" indicates the existence of a defect and "$X = 0$" the absence of a defect. The interest of this formulation lies in the fact that it offers a clear definition of the Probability of False Alarm (PFA):

$$\text{PFA}(X) = P\left[d(X) = 1 | X = 0\right]$$
[14.16]

If the defect is an event with non-discrete values where the distribution of the signal is known, and the distribution of the noise is known, the theory of detection leads to the following definitions of the two probabilities, PoD and PFA:

$$\text{PoD} = \int_{a_d}^{+\infty} f_{SN}\left(\hat{a}\right) d\hat{a} \quad ; \quad \text{PFA} = \int_{a_d}^{+\infty} f_N(\eta)\, d\eta \qquad [14.17]$$

where f_{SN} and f_N indicate, respectively, the probability densities of the variables "signal + noise" and "noise". We can note that the probability density of the noise can be defined by the probability conditioned by the measured value. We shall not detail these considerations because it is extremely difficult to prove and even to quantify them. We can simply note that, physically, for many measurement devices, the operator can tune the signal gain more and more finely, if defects are not detected with the original settings. In this case, the noise evolves with the adjustment and consequently with the defect that we are measuring. The formulas PoD and PFA can be modified to include this information in the conditional probabilities.

For a given size or class of measured defect, we can plot the curve relating to the points with coordinates (PFA; PoD), by modifying the parameters affecting the measurements (according to the case concerned, the parameters can be device adjustment, visibility, operator experience, etc.). This curve, Figure 14.2, is obtained, in a continuous form, by varying the threshold a_d; the curve is called the curve of Receiver Operating Characteristics, or simply the "ROC curve".

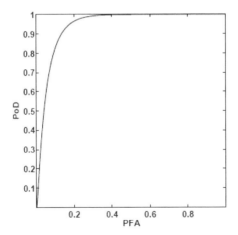

Figure 14.2. *Receiver Operating Characteristic (ROC) curve: evolution of the probability of detection (PoD) versus the probability of false alarm (PFA) [ROU 01]*

Note that, in case of inspections under severe *in situ* conditions, such as in high mountains, offshore platforms and marine structures, the performance of measurement devices is strongly affected (be agitation of waves and storms, visibility, temperature, experience and state of fatigue of divers, quality of the link with platform supervisor, etc.). Campaigns of inter-calibration of type – InterCalibration of NDT for Offshore Structures (ICON) – become necessary, by which we measure, for each class of defect (size and typology), the numbers of good and bad detections, and calculate the observed probabilities corresponding to the two cases:

$$P_b(c) = \{p_b(c), p_r(c)\} \text{ where } \begin{cases} p_b(c) = \dfrac{n_b(c)}{n_b(c) + n_n(c)} = \dfrac{n_b(c)}{n_1(c)} \\ \\ p_r(c) = \dfrac{n_r(c)}{n_f(c) + n_r(c)} = \dfrac{n_r(c)}{n_2(c)} \end{cases} \qquad [14.18]$$

$$P_F(c) = \{p_F(c), p_n(c)\} \text{ where } \begin{cases} p_F(c) = \dfrac{n_F(c)}{n_2(c)} \\ \\ p_n(c) = \dfrac{n_n(c)}{n_1(c)} \end{cases} \qquad [14.19]$$

where $n_b(c)$, $n_F(c)$, $n_n(c)$ and $n_r(c)$ are, respectively, the number of existing and detected defects, the number of non-existing and detected defects, the number of existing and undetected defects, and finally the number of non-existing and undetected defects. According to these definitions, $p_F(c)$ is the PFA and $p_b(c)$ is the PoD. We can, depending on the considered class of defects, build discrete ROC curves.

14.4.2.2. *New concepts for decision-making*

For a structure manager, the questions are often different and new probabilities have to be introduced [ROU 03]. In fact, when using Bayesian modeling, we have to define the conditional probabilities associated to the following events:

– E_1: non-existence of a crack knowing that a crack is not detected;

$$P[E_1] = P_1 = P[X = 0 \mid d(X) = 0]$$

– E_2: non-existence of a crack knowing that a crack is detected;

$$P[E_2] = P_2 = \Pr[X = 0 \mid d(X) = 1]$$

– E_3: existence of a crack knowing that a crack is not detected;

$$P[E_3] = P_3 = \Pr[X = 1 \mid d(X) = 0]$$

– E_4: existence of a crack knowing that a crack is detected;

$$P[E_4] = P_4 = \Pr\left[X = 1 \mid d(X) = 1\right]$$

Some of these events are complementary and we can deduce the relationship between their probabilities:

$$P_1 + P_3 = 1 \; ; \; P_2 + P_4 = 1 \tag{14.20}$$

We can write these probabilities in terms of PoD and PFA to find:

$$P_1 = \frac{(1 - \mathrm{PFA}(X))(1 - \mathrm{PCE}(X))}{(1 - \mathrm{PoD}(X))\,\mathrm{PCE}(X) + (1 - \mathrm{PFA}(X))(1 - \mathrm{PCE}(X))}$$

$$P_2 = \frac{\mathrm{PFA}(X)(1 - \mathrm{PCE}(X))}{\mathrm{PoD}(X)\,\mathrm{PCE}(X) + \mathrm{PFA}(X)(1 - \mathrm{PCE}(X))}$$

$$P_3 = \frac{(1 - \mathrm{PoD}(X))\,\mathrm{PCE}(X)}{(1 - \mathrm{PoD}(X))\,\mathrm{PCE}(X) + (1 - \mathrm{PFA}(X))(1 - \mathrm{PCE}(X))}$$

$$P_4 = \frac{\mathrm{PoD}(X)\,\mathrm{PCE}(X)}{\mathrm{PoD}(X)\,\mathrm{PCE}(X) + \mathrm{PFA}(X)(1 - \mathrm{PCE}(X))}$$

These equations introduce a new measure of probability: the Probability of Crack Existence (PCE), so named because these definitions were initially developed for the detection of cracks in the oil structure industry. The presence of only probabilities PoD and PFA in the same decision scheme is not, therefore, satisfactory. Moreover, considering only the PoD is equivalent to considering that PoD = P(d(X) = 1). This implies that the two conditions: {PCP = 1 ; PFA = 0}, are satisfied, which are strong assumptions. Parametric studies can thus be performed, in order to identify (for example) the importance of the PFA. Hence, the information transfer during inspection can be drawn as indicated in Figure 14.2, where $\mathfrak{I}_1$ is an unknown function and $\mathfrak{I}_2$ is described by the nonlinear equations above.

We note that the laboratory generated Probability of Detection (PoD) is discontinuous by class of defects, while the decision-maker needs continuous information scales, integrable and differentiable for a numerical analysis.

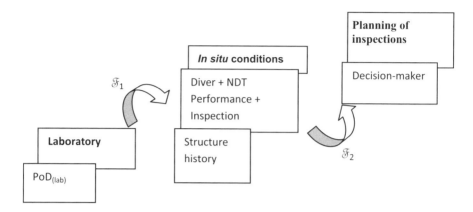

Figure 14.3. *Inspection information transfer in the decision process*

Figure 14.4 depicts the evolution of probabilities P_2 and P_3 for the Probabilities of Crack Existence (PCE) varying from 0.1 to 0.5. The ROC curves are therefore defined by projections in the plane (PoD, PFA) of the inspection operating curves on these surfaces. Three of the ROC curves are illustrated in Figure 14.4.

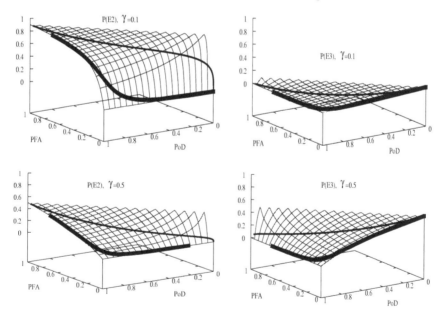

Figure 14.4. *Variations of P_2 (left) and P_3 (right) in the plane (PoD; PFA) for the probabilities of crack existence PCE=γ= 0.1 (top figures) and 0.5 (bottom figures)*

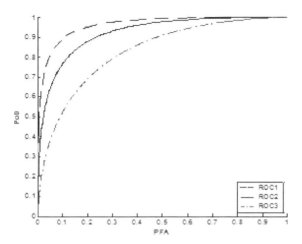

Figure 14.5. *Evolution of the Probability of Detection (PoD) as a function of the probability of false alarm (PFA): Receiver Operating Characteristic (ROC) curves for the Probabilities of Crack Existence (PCE) under various conditional probabilities P_i*

From the curves in Figure 14.5, it can be observed that ROC curves are highly sensitive to the variations of PCE and to the studied conditional probabilities P_i.

14.5. Structures with large lifetimes

For structures with large lifetimes, such as civil engineering structures and infrastructures, it is necessary to take into account the evolution of monetary values, which is performed by the mean of discount functions, including interest and inflation rates. Moreover, the assumption of infinite horizon cannot usually be allowed, as the number of actions is often limited during the lifetime of the structure. In this case, the total cost concerning the whole lifetime of the structure should be considered, and should include discount effects. When failure occurs between two inspections at times t_{i-1} and t_i, the expected failure cost is written as:

$$C_F(N_{INS}) = \sum_{i=1}^{N_{INS}} (F(t_i) - F(t_{i-1})) \frac{C_f(t_i)}{(1+r)^{t_i}} \qquad [14.21]$$

where $C_f(t_i)$ is the cost of failure consequences. Moreover, for a number of inspections N_{INS}, the expression of the total inspection cost is written:

$$C_{INS}(N_{INS},q) = \sum_{i=1}^{N_{INS}} (1 - F(t_i)) \frac{C_{ins}^i(q)}{(1+r)^{t_i}}$$ [14.22]

where $C_{ins}^i(q)$ is the cost of the ith inspection which depends on type and the quality q, $F(t_i)$ is the cumulated failure probability at the ith inspection and r is the discount rate (often considered between 0.01 and 0.05, and up to 0.09 in the nuclear industry). At the end of each inspection, we can associate the Probability of Detection $PoD(t_i)$ and the Probability of False Alarm $PFA(t_i)$; a decision should then be taken regarding the system repair, taking account for $PoD(t_i)$ and $PFA(t_i)$. This decision is generally based on admissible reliability levels. The repair and replacement cost C_{REP} depends on the nature and number of actions N_{INS} to be performed:

$$C_{REP}(N_{INS},q) = \sum_{i=1}^{N_{INS}} P_{REP}(t_i) \frac{C_{rep}^i(q)}{(1+r)^{t_i}}$$ [14.23]

where $C_{rep}^{i'}(q)$ is the replacement cost at the ith inspection and $P_{REP}(t_i)$ is the corresponding repair probability.

14.6. Criteria for choosing a maintenance policy

Maintenance policies can be based on various criteria to define optimal strategy. Garabatov & Guedes Soares [GAR 01] compared strategies based on the following criteria:

– pure economic criterion: the time intervals between inspections and replacements are defined by optimal cost of maintenance without constraints on the required reliability level;

– economic criterion with minimal interval: in order to avoid closely scheduled operations, a constraint on the minimum time interval between successive operations is introduced in the cost optimization problem;

– pure operational criterion: for a better management of the system and its availability, a constant time interval is often adopted for maintenance operations; the choice of this interval is based on a minimization of the total maintenance cost;

– pure reliability criterion: the time interval is determined by the time at which the system reliability reaches the minimum acceptable level; due to system degradation, the time intervals vary along the lifetime of the structure;

– reliability criterion based on inspection quality: in this case, the inspection/replacement intervals are regular, but the quality of the operation is adjusted such that minimum reliability is ensured over the whole lifetime.

In general, the purely economic criterion leads to a large reduction of costs, but implies frequent maintenance actions. The choice of a specific policy strongly depends on the nature of the system and the failure consequences. A reliability criterion with consideration of maintenance quality seems to be a reasonable compromise to reduce costs, while ensuring appropriate reliability levels.

14.7. Example of a corroded steel pipeline

To illustrate some of the above concepts, consider a simple example of a steel pipeline subject to corrosion. The system variables and their distribution parameters are given in Table 14.1 (for simplicity, all the probability distributions are considered as normal). In this example, the tube wall thickness loss is given by the corrosion law of type kt^n for $t > 1$, where t is the time in years, and k and n are the parameters of the corrosion model.

By considering the safety margin corresponding to the material strength regarding hoop stress, the reliability index is found to be 3.904 (the mean of the margin is 4.5 and its standard deviation is 1.153). In the corroded state, the safety margin, the reliability index and the failure probability are given by:

Safety margin: $G(t) = f_Y(e_0 - kt^n) - p\,r$

Reliability index: $\beta(t) = \dfrac{(e_0 - kt^n)\,m_{f_y} - r\,m_p}{\sqrt{\left((e_0 - kt^n)\,\sigma_{f_y}\right)^2 + \left(r\,\sigma_p\right)^2}}$

Failure probability: $P_f(t) = \Phi(-\beta(t))$

Variable	Symbol	Mean	Standard deviation	Units	
Internal pressure	p	4.5	0.9	MPa	
Yield stress	fy	360	28.8	MPa	
Mean radius	r	200	-	mm	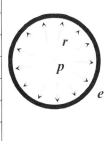
Thickness	e_0	5	-	mm	
Parameter 1	k	0.005	-	mm/yr$^{1.4}$	
Parameter 2	n	1.4	-	-	

Table 14.1. *Geometrical and mechanical characteristics of the pipeline*

Figure 14.6 depicts the evolution of the reliability index β as a function of the age of the structure t. Table 14.2 indicates various costs involved along the lifetime of the pipe.

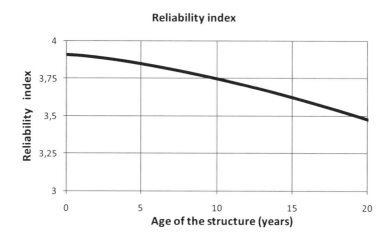

Figure 14.6. *Reliability index as function of the pipe age*

Initial cost of manufacturing and installation	$C_0 = 600$ k€
Failure cost	$C_F = 30000$ k€
Perfect preventive maintenance cost	$C_{PM} = 20$ k€
Imperfect preventive maintenance cost	$C_{IM} = 10$ k€/mm

Table 14.2. *Costs involved during the pipe's life*

In this example, perfect maintenance corresponds to replacement of the pipe by a new one, and imperfect maintenance consists of applying a coating with cost equal to 10 k€ per mm of additional thickness. Without maintenance, the total cost of the pipe is composed of the initial and failure costs. The expected total cost is plotted in Figure 14.7 as a function of the pipe age, where a minimum is observed at 41 years, corresponding to its economic lifetime without maintenance.

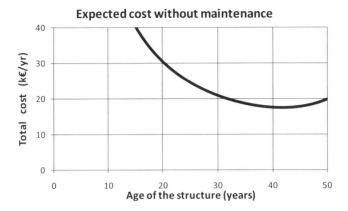

Figure 14.7. *Total cost without maintenance*

Figures 14.8 and 14.9 depict the costs per unit time in the case of perfect and imperfect maintenance, respectively. In the case of perfect maintenance, optimal maintenance is located at 23 years with a total cost of 1.34 k€/yr. In the case of imperfect maintenance, we have chosen to add 0.3 mm of coating, representing a repair cost of 3 k€. The optimum is located at 16 years with a total cost of 0.49 k€/yr. It is important to note that these values are based on the assumption of an infinite horizon. It can easily be demonstrated that this assumption does not apply in the case of imperfect maintenance, as the maintenance cycles are not identical. In other words, imperfect maintenance is only valid for the first cycle.

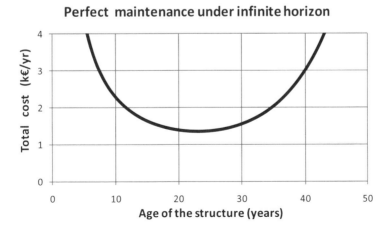

Figure 14.8. *Perfect maintenance cost according to time interval between operations*

By considering the case of a finite horizon, the curves in Figures 14.10 and 14.11 show the evolution of the failure probabilities associated with different perfect and imperfect maintenance policies, respectively. In this case, we have to calculate the total cost over the service life, which is taken here as 50 years. In the case of perfect maintenance, Table 14.3 indicates various policies and their corresponding costs.

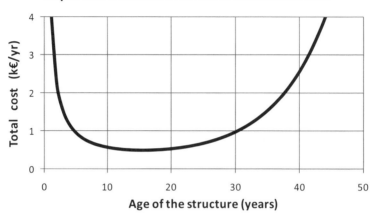

Figure 14.9. *Imperfect maintenance cost according to time interval between operations*

Perfect maintenance				
Policy	**Maintenance time (years)**	**Total cost (k€)**	**Cost per unit time (k€/yr)**	**Remarks**
One action (infinite horizon)	23	48.9	0.978	Interval obtained with the assumption of infinite horizon
One action (finite horizon)	25	47.9	0.958	Interval at 50% of the service life, in order to balance the failure probabilities in the two intervals
Two actions	16 and 32	55.7	1.114	Low degradation levels, but high cost of maintenance

Table 14.3. *Maintenance costs in terms of the number of actions and type of horizon*

As the two cycles are supposed to start with a new structure, the failure costs are minimized when the two cycles are identical (i.e. a maintenance operation at 25 years), which explains why the consideration of finite horizon allows us to reduce the total cost. The application of two maintenance actions leads to a significant increase in the maintenance costs, which is not recovered by the benefits of reducing the failure costs.

Figure 14.10. *Evolution of the failure probability (perfect maintenance)*

In the case of imperfect maintenance, the assumption of an infinite horizon allows us to optimize the first cycle, but the increase of the failure probability at the end of the lifetime (i.e. at 50 years) leads to very large failure costs. When the maintenance intervals are chosen to balance the failure probabilities by using 0.5 mm of coating, at 15 and 35 years, we obtain a total costs of 18.9 k€ instead of 44 k€. The same strategy is applied with four operations, leading to a higher cost of 21.2 k€.

Imperfect maintenance				
Policy	**Maintenance times (years)**	**Total cost (k€)**	**Cost per unit time (k€/yr)**	**Remarks**
Two actions (coating = 0.3mm)	16 and 32	44.7	0.894	Intervals obtained by the infinite horizon assumption
Three actions (coating = 0.5mm)	15 and 35	18.9	0.37	Intervals that balance the maximum failure probabilities
Four actions (coating = 0.28mm)	10, 20, 30 and 40	21.2	0.424	Low degradation levels

Table 14.4. *Maintenance costs in terms of the number of actions and type of horizon*

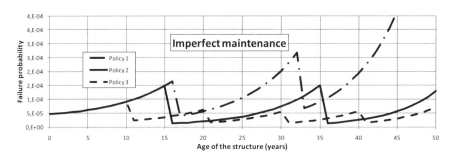

Figure 14.11. *Evolution of the failure probability (imperfect maintenance)*

14.8. Conclusion

This chapter has introduced several types of reliability-based maintenance models. The main difficulty lies in the estimation of direct and indirect costs of failure, especially when immaterial losses are involved (i.e. human lives, public relations effects, etc.). The formulation of the maintenance cost becomes more difficult when multi-component systems are considered, as economic and stochastic interactions make the analysis very complex. Interested readers can consult the specialized literature, such as [CRE 03], dedicated to the management of infrastructures by considering inspection tool performance, determination of degradation laws, reliability assessment and the choice of maintenance actions.

14.9. Bibliography

[BRE 09] BREYSSE D., ELACHACHI S.M., SHEILS E., SCHOEFS F., O'CONNOR A., "Life cycle cost analysis of ageing structural components based on non destructive condition assessment", *Australian Journal of Structural Engineering*, 9:1, p. 55-66, 2009.

[CRE 03] CREMONA C. (ed.), *Application des notions de fiabilité à la gestion des ouvrages existants*, Edition Presses Ponts et Chaussées, Paris, France, 2003.

[FAB 02a] FABER M.H. "RBI: An Introduction", *Structural Engineering International*, 3/2002, p. 187-194, 2002.

[FAB 02b] FABER M.H., SORENSEN J.D., "Indicators for inspection and maintenance planning of concrete structures", *Journal of Structural Safety*, 24, 2002.

[GAR 01] GARBATOV Y., GUEDES SOARES C., "Cost and reliability based strategies for fatigue maintenance planning of floating structures", *Reliability Engineering and System Safety*, 73, p. 293-301, 2001.

[MAD 87] MADSEN H., SKJONG R., TALLIN A. and KIRKEMO F., "Probabilistic fatigue crack growth analysis of offshore structures, with reliability updating through inspection", *Marine Structural Reliability Symposium*, p. 45-55, Arlington, Virginia, USA.

[MOA 97] MOAN T., VÅRDAL O.T., HELLEVIG N.C., SKJOLDLI, "In-Service Observations of Cracks In North Sea Jackets. A Study on Initial Crack Depth and POD values", *Proceeding of 16th International Conference on Offshore Mechanics and Arctic Engineering, (O.M.A.E'97)*, Vol. II Safety and Reliability, p. 189-197, ASME editor, 1997.

[MOA 98] MOAN T., SONG R., "Implication of inspection updating on system fatigue reliability of offshore structures", *Proceeding of the 17th International Conference on Offshore Mechanics and Arctic Engineering*, no. 1214, ASME editor, 1998.

[MOA 99] MOAN T., JOHANNESEN J.M., VÅRDAL O.T., "Probabilistic inspection planning of jacket structures. In: Offshore Technology Conference Proceedings", *Offshore Technical Conference*, paper no. 10848, Houston, TX, USA, 1999.

[ROU 01] ROUHAN, A., Structural integrity evaluation of existing offshore platforms based on inspection data, PhD, thesis, University of Nantes, France, 2001.

[ROU 03] ROUHAN, A., SCHOEFS, F. "Probabilistic modelling of inspections results for offshore structures", *Structural Safety*, 25, p. 379-399, 2003.

[SCH 09] SCHOEFS F., "Risk analysis of structures in presence of stochastic fields of deterioration: coupling of inspection and structural reliability", *Australian Journal of Structural Engineering*, Special Issue - Disaster & Hazard Mitigation, Vol. 9, no. 1, 2009.

[SHE 10] SHEILS E., O'CONNOR A., BREYSSE D., SCHOEFS F., YOTTE S., "Development of a two-stage inspection process for the assessment of deteriorating bridge structures", *Reliability Engineering and System Safety*, 95:3, p. 182-194, 2010.

Chapter 15

Practical Aspects: Industrial Implementation and Limitations in a Multi-criteria Context

15.1. Introduction

If the consideration of a purely economic criterion can be the optimal solution for a simple and well-known system, in practice, things are rarely as easy as a result of interactions with other systems. This is particularly the case for any complex industrial system.

To illustrate this, we can take the example of optimizing a system of waterworks in a city: this system includes elements (pipes) of different ages and different types depending on the development of the city and neighborhoods, different loads (low or high road traffic), and a different environment (more or less aggressive soil in the case of metallic pipes for example), to illustrate only a few parameters among many others.

The economic optimization of the maintenance of such a system is theoretically possible, provided sufficient information is available. Such information concerns the physical behavior, economic valuation, the failure rate or the kinetics of degradation by the factors mentioned above. However, this system also interacts with other systems in the city – sewage, electricity, and gas systems for example – which theoretically have their own optimal maintenance frequency.

We must therefore consider economic optimization from a global point of view at the city scale or (more practically) the neighbourhood scale to avoid digging

---

Chapter written by Franck Schoefs and Bruno Capra.

through a road that was renovated a year ago in order to replace some pipes. In addition, other parameters that are not purely technical come into play: noise, traffic problems, serviceability or, more generally, anything that is related to users can also impact the optimization system if social criteria are integrated.

Finally, other even more uncertain aspects are to be considered: investment opportunities (loss of revenue in crisis period), the change of development strategy related to a change of government, etc.

This small example shows that it rapidly becomes very difficult to optimize a complex multi-criteria system. It is even possible for contradictory stakes to appear that must then be studied in different ways: global optimization weighted in accordance with the stakes considered, or separation and optimization of some main stakes. In summary, any optimization process should be understood in the context of a very specific reference framework that takes into account the most important stakes for the operator.

An important aspect in the development of a methodology for industrial optimization of maintenance concerns, as outlined in the previous chapters, is the quantity and quality of data. Frequently, if we consider the case of large industrial assets, management practice is to follow the hierarchy of needs presented in Figure 15.1.

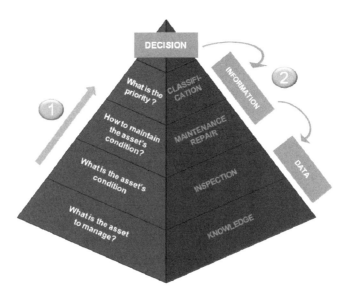

Figure 15.1. *Pyramid of needs for the management of industrial assets (from [BOE 09])*

Traditionally, the manager of an industrial asset follows the pyramid from the base to the top (arrow 1, Figure 15.1):

– knowledge: what is the asset to be managed? The answer to this question is not always evident for an old structure for which the history could have been lost with time (fire, war, relocation, etc.). It is necessary to clearly define the perimeter of the asset to study before any other action is taken;

– survey: what is the condition of the asset? This question is often perceived as the most important because it is prior to any optimization phase. However, it does not always need to be answered fully (100% of the known assets) as we shall see later. This step allows us to define the state of degradation of the progressively ageing asset or to identify the current service life of elements;

– maintenance/repair: how will we maintain or recover the asset? To optimize the overall cost of maintenance actions, it is necessary to know the different possibilities (options) for maintaining or repairing the system, and the individual costs associated. In the case of infrastructure, for the same failure mode, different levels of maintenance actions may be available. These may have varying costs, but also varying efficiency. In the case of a cost/benefit approach, this element should not be forgotten when deciding on optimal maintenance actions;

– ranking: what actions should be taken first? This phase requires the definition of one or more limit states and corresponding thresholds for optimization. In the context of industrial infrastructure, criteria related to safety, availability and cost are commonly considered;

– decision: the final stage of the process. All previous steps provide key elements for decision support in relation to the various stakes considered.

This classic *bottom-up* approach is often seen as applicable only if the state of the system is perfectly known, which is not necessarily the case. Therefore, if ageing is controlled, the different steps generally involve the prior knowledge of the whole system before continuing the process (inspection). This is disadvantageous because it requires potentially significant investment and time before the definition of the maintenance master plan.

Approaches for risk-based maintenance, such as Failure Mode, Effects and Criticality Analysis (FMECA, see Part 1) or Reliability Centered Maintenance (RCM, see Chapter 13) for example, can address the problem of optimizing maintenance without first having complete knowledge of the system (arrow 2, Figure 15.1). Given the priorities in terms of maintenance policy, the various stakes considered, and partial knowledge of the asset, it is possible to define an action plan which recommends improving knowledge of the asset or monitoring some targeted elements only, rather than the whole system.

Considering the previous example, if we consider a sewer network, it is not necessary to know the current status of the entire system by inspection, which would be very expensive and time consuming, before taking action. A preliminary risk analysis is used to define the highest risk areas, with some degree of uncertainty given partial existing data, and then enables the definition of the main areas to inspect in order to refine the analysis and optimize the corresponding maintenance actions.

In summary, the industrial application of maintenance optimization methods based on a single economic optimum is difficult to implement from a practical point of view: given the uncertainties and complex phenomena involved, it is often more convenient to reason with respect to different scenarios that can encompass various alternatives. The following paragraph gives some examples of maintenance optimization in different industrial contexts.

15.2. Motorway concession with high performance requirements

15.2.1. *Background and stakes*

This study concerns the concession of a European motorway section for which the contractor must provide the entire integrated project design, construction, financing, maintenance and operation over the next thirty years. To ensure network availability, security and quality for users, a penalty system was introduced at the initiative of the public authority. Given the significant penalties for non-compliance with the criteria of availability, the contractor faces a high risk potential associated with unplanned outages, despite the precautions taken to ensure reliable operation. Therefore, there is a strong challenge for the contractor, during the call for tender, to fund the cost of risk associated with unavailability of the motorway network.

Unavailability can be of two different types: planned (for maintenance intervention) or unplanned. The challenge is to assess the residual risk associated with these unplanned outages. In particular, the following steps must be undertaken:

– identify and assess the risk:

   - frequency of occurrence: identify the ageing mechanisms that might call into question the availability criteria, evaluate their kinetics, and determine the associated probability of failure;

   - severity: identify the different scenarios of penalties;

– monetize the risk: financially quantify the amount of risk;

– characterize the quantitative risk over 30 years. This enables us to know how and when to plan for risks.

In a second phase, the objective is to optimize the overall forecast operating costs over the operating period. These costs include:

– construction costs, depending on the chosen design;

– maintenance costs over the entire concession time;

– operating costs over the entire concession time; and

– the cost of risk associated with unavailability.

The results of the study lead to a connection between a financially quantified residual risk, relative to unplanned unavailability, and different scenarios related to the quality of design and maintenance levels implemented. The more robust the design choices are, and the greater the level of maintenance, the lower the residual risk, and vice versa. The final design and maintenance choices are taken with a view to achieving the optimum scenario to minimize the total cost of the project.

The asset studied consists of a 40-year old highway network of about 15 km that connects three highways and crosses a canal through an underwater tunnel of about 600 m length. The existing tunnel has two partitions of one-way traffic made of two lanes in each direction. The decision was taken to build a new tunnel with two partitions parallel to the former (see Figure 15.2).

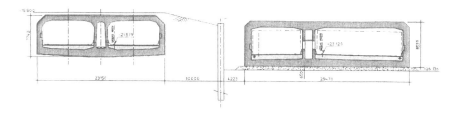

Figure 15.2. *Studied system: existing and new tunnel (from [AUG 09])*

To minimize the possibility that the tunnel undergoes multiple unplanned maintenance periods, the contract for construction and operation includes a penalty clause related to unavailability but also some quality and comfort criteria. About thirty operational requirements were defined in the technical specifications, from the pavement quality (roughness, drainage, etc.) to the proper operation of safety systems (lighting, ventilation, light signals, etc.). For each of these criteria, beyond a certain level of system failure (such as the number of lights out of service) or degradation of the roadway (such as time of drainage of rain water), maintenance is imposed and the section over which the failure occurs is considered to be unavailable, triggering penalties.

The severity of the penalty applied depends on several factors, including the duration of the unavailability of the section concerned, the number of lanes closed, the time at which the unavailability occurs (peak, off peak, night, etc.). These penalties can quickly reach several million euros.

A possible preventative measure consists of anticipating failures and planning maintenance actions before they are imposed. The system operator is allowed 35 nights per year of downtime for maintenance without penalty. The challenge is to find the optimum maintenance program to balance the anticipation of works, which helps to avoid unplanned outages, and postponement, avoiding the proliferation of maintenance actions.

15.2.2. *Methodology*

A preliminary analysis showed the requirements of the most critical operation:

– pavement roughness;

– accumulation of water on the roadway;

– operation of the ventilation system in tunnels;

– operation of traffic detectors;

– operation of the monitoring and management system (control room).

The approach adopted for the study of unavailability related to a problem of pavement roughness and to failure of the ventilation is presented below. In general, the probability of failure over time was calculated initially on the basis of available feedback. The probability of failure was then used to determine the cumulative probability of a failure during the concession period of 30 years.

15.2.2.1. *Roughness of the pavement*

The roughness of a roadway increases during the first year of operation due to the gradual erosion of the binder around the aggregates, which then become sharper. The initial friction coefficient may therefore be relatively low and the probability of non-compliance with the roughness requirement is then non-zero. When the surface binder has disappeared, the coefficient of friction decreases due to aggregate polishing and stabilizes after about 5 years. The probability of failure increases during this period until the road surface is renovated ($\approx$ 10 to 13 years).

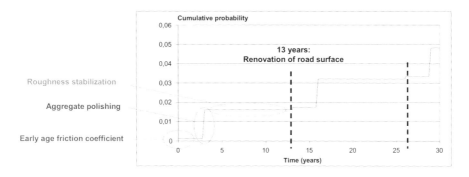

Figure 15.3. *Cumulative probability of unavailability
due to the roughness of the pavement [AUG 09]*

The cumulative penalties cost, shown in Figure 15.4, is based on the number of lanes that would be impacted by the unavailability and takes into account the specificities of different sections of the roadway.

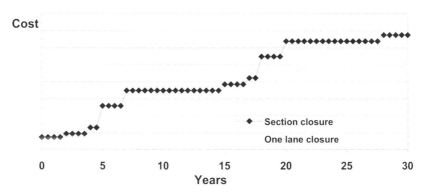

Figure 15.4. *Cumulative penalty costs for unavailability
due to non-compliance with roughness requirements*

15.2.2.2. *Ventilation system in tunnels*

For these kinds of systems, the probability of failure was determined according to the manufacturers' data using a Weibull type law. The probability distribution for each component was fitted on the basis of expected lifetimes (frequency of planned replacement), the mean time between failures (MTBF) provided by manufacturers, and usual lifetimes during which a failure is very unlikely to occur.

For ventilators, the combination of the probability of failure of the devices themselves (wear of propellers) and electronic and electrical components is considered.

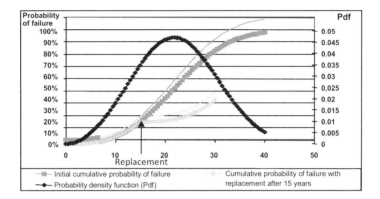

Figure 15.5. *Probability of failure of a ventilator*
(no electronic or electrical device considered)

To calculate the probable cost of penalties, several design choices were available including possible redundancy of ventilators (Figure 15.6).

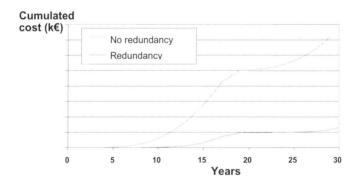

Figure 15.6. *Cumulative cost of residual risk of unavailability due to failure*
of ventilators depending on the design choices

15.2.3. *Results*

The approach taken has helped to assign a financial cost to the risk of unavailability associated with performance requirements, in terms of quality and comfort for users, throughout the concession period. For each requirement, the accumulation of penalties has been estimated to determine the full cost penalties associated with the residual risk of unavailability considering the planned maintenance (Figure 15.7).

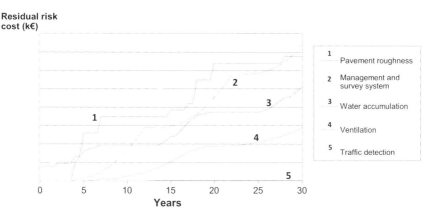

Figure 15.7. *Cost of residual risk of unavailability for different requirements*

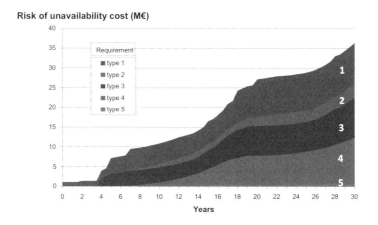

Figure 15.8. *Cost of residual risk associated with unplanned unavailabilities of the road*

The calculation of the total cost of penalties for unavailability was used to validate design choices: between an optimum design resulting in less downtime but with higher initial cost (e.g. redundant systems) and the gain provided by this solution in terms of cost of risk associated with these penalties. Similarly, the residual cost of risk was also calculated by considering different maintenance intervals.

At the end of the study, the contractor obtained a probabilistic estimate of the penalties they would have to pay with costs distributed over the 30 years of operation. This allows identification of periods where preventive maintenance would have to be done to avoid "peaks" of failure. The results obtained lead to a quantitative view of the residual risk associated with unplanned unavailability of the motorway network. In addition, knowledge of the potential penalty distribution for the next 30 years enables the optimized programming of budgets.

The resulting value of around €35 million over 30 years (averaging close to €1 million per year) is significant compared to the magnitude of the total project cost of approximately €500 million. This result is related to the very high penalties for non-compliance with the availability criteria. The benefit of such a comprehensive approach to infrastructure lifecycle management applies to the building owner, prime contractor, contractor or operator. In all cases, the method has many advantages:

– clear definition of performance objectives, criteria and associated indicators;

– clear allocation of responsibilities to each party;

– transparency towards the identified risks and their methods of treatment (who is responsible? What actions should be implemented?)

– argued defense of budgets;

– optimization of the total costs of infrastructure projects in relation to the entire life cycle;

– consideration and preservation of the specific stakes of each party;

– risk-based monitoring and maintenance of infrastructure driven by the dual consideration of ageing mechanisms and impacted stakes, according to the parties and the periods considered.

This example shows the importance of proper consideration of short-term choices (design, construction methods) and short-term stakes (from day-to-day availability), and the long-term implications of these choices (maintenance costs and condition of the structure at the end of the concession contract) as well as long-term challenges (asset condition, sustainable management).

15.3. Ageing of civil engineering structures: using field data to update predictions

15.3.1. *Background and stakes*

Risk management of ageing assets, particularly civil engineering infrastructure, is an important issue for the future. The technical challenge is not only to build new infrastructure but also to maintain what already exists because the economic stakes are considerable. The prediction of ageing through modeling allows management of the risks associated with the expected service life of infrastructure. The use of updating techniques enables the best use to be made of monitoring instrumentation and inspection data in order to define optimized maintenance strategies. Benefits for the asset owners include better risk management and savings on maintenance budgets. To anticipate and optimize these costs, it is necessary to use ageing models which are as representative as possible of the real physical asset, and input data which is as reliable as possible.

In the field, some characteristic ageing parameters can be measured more or less accurately. Bayesian approaches (Chapter 11, section 11.3) are a technique well adapted to this type of problem to update the ageing prognosis. This kind of prediction is generally based on the use of physical models where various parameters of the models may be uncertain. This uncertainty can be divided into two types:

– the inherent variability of these parameters, such as the compressive strength of concrete which will vary within a single mix because of the heterogeneity of the material;

– uncertainty of measurement linked to the device used to quantify the parameter or a lack of knowledge about these parameters.

In practice, some physical material characteristics can be measured, as well as the consequences of ageing. However, like any measurement, these parameters have some uncertainty. In addition, there are generally relatively few measurements made because the techniques are expensive and of varying accuracy. It is therefore important to use them effectively with proper data processing. Bayesian methods can be used to process data and update the ageing prognosis.

15.3.2. *Corrosion risk of a cooling tower*

The physical phenomenon considered here is the carbonation of concrete. Carbon dioxide from the atmosphere penetrates concrete due to its porosity and dissolves in the pore solution. The carbonic acid formed then reduces the pH of the concrete. If the carbonation front reaches the rebars, they are no longer protected by the highly basic pH of sound concrete and will therefore gradually be affected by corrosion. Corrosion is the most common cause of deterioration of reinforced concrete structures and also the most expensive to repair. It is therefore crucial to assess the risk of corrosion to optimize the maintenance of structures that may be susceptible to this pathology [ELL 95].

The example presented here concerns a nuclear plant's cooling tower [CAP 07]. Considering the dimensions of the structure (height greater than 100 m) and the environment to which it is subject over time (wetting/drying cycles, temperature gradients, moisture, etc.) material properties and solicitations vary in time and space. During an inspection campaign after 25 years of operation, samples of concrete were taken from the structure. The values of compressive strength and carbonation depth were measured and are summarized in Table 15.1.

	Mean value	Standard deviation	Number of measurement
Compressive strength Rc (MPa)	48.8	4.7	15
Carbonatation depth X (mm)	6.3	2.8	9

Table 15.1. *Statistics derived from the experimental campaign*

In parallel, measurements of concrete cover (d) were also carried out (see Figure 15.9).

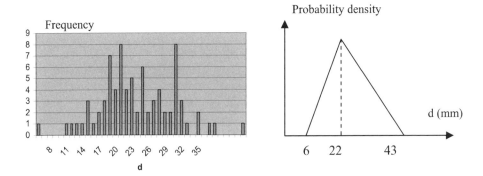

Figure 15.9. *Experimental data and modeling of the statistical distribution of concrete cover (d)*

An exhaustive visual inspection of the structure was carried out to measure the total length of the unprotected steel reinforcement. Rebar corrosion causes cracking of the concrete cover because the volume occupied by the corroded steel is larger than that occupied by sound steel. The inspection determined that the total length of exposed steel was 115 m out of a total of about 500 km of reinforcing steel in the whole outer wall of the cooling tower. This means the order of magnitude of the proportion of exposed rebars to the cumulated length of steel is 10^{-4}. We can therefore deduce that the real proportion of corroded reinforcement at 25 years is greater than this value of 10^{-4} because a certain amount of steel is corroded without external signs, i.e. having yet reached the pressure needed to produce cracking or spalling of the concrete cover.

15.3.3. *Bayesian updating*

To cope with uncertainty about the parameters of the carbonation model used to describe the evolution of phenomena and reflect data from the experimental field, an updating technique based on a Bayesian network was developed using a Monte Carlo Markov Chain (MCMC) approach, given the available measurements at different nodes of the network (Figure 15.10).

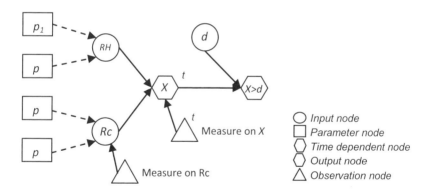

Figure 15.10. *Bayesian network used for the carbonation model (RH: relative humidity)*

The graph of this Bayesian network represents a carbonation model developed to calculate the probability of corrosion initiation of steel in reinforced concrete considering existing measurements of *Rc* (concrete compressive strength) and *X* (depth of carbonation). Nodes linked by solid lines represent physically dependent random or deterministic variables; the dotted lines represent the links between random variables and parameters. The model inputs, i.e. *RH* (relative humidity) and

Rc (compressive strength), are modeled by conditional laws depending on a number of parameters $p_1, ..., p_4$. Each of these parameters is modeled by a random variable. Each measurement is modeled by a random variable depending on the input it measures. In addition, the output variables X (depth of carbonation) and $X > d$ (limit state function: depth of carbonation X greater than concrete cover d = initiation of corrosion), and measurement variables of X, depend on time t.

To calculate the probability of corrosion initiation over time, a Beta distribution law has been postulated for the variability of the compressive strength according to data from Table 15.1 and the fact that, physically, this parameter is bounded by minimum and maximum values (estimated from Table 15.1).

A deterministic approach is not capable of representing the apparent ratio of corroded steel observed on the structure (approximately 10^{-4} after 25 years) as it does not take into account the variability that affects the corrosion process. The probability of corrosion initiation obtained by using the carbonation model in the probabilistic approach gives a value of about 10^{-2} after 25 years. This value is higher than the observed ratio because it does not correspond to the same indicator: the corrosion process has to be sufficiently advanced (a certain amount of corrosion products i.e. of steel loss) before cracks appear at the surface of the structure. The order of magnitude predicted here is a good estimate if we consider that cracks appear for a loss of steel cross-section of about 100 to 200 μm.

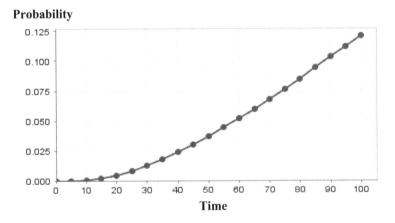

Figure 15.11. *Probability of corrosion initiation over time, P(X(t) > d)*

A Bayesian network is a probabilistic graphical model that can acquire and use information to update system knowledge. In this example, forecasting the evolution of carbonation depth based on experimental data collected after 25 years of operation and estimated initial data has been updated (Figure 15.12). Compared to the initial prediction of carbonation depth, it can be seen that feeding field data into the updating process has led to a decrease of the mean value but also the uncertainty associated to the carbonation front (time fractiles at 5 and 95% are closer to the mean).

Carbonation front X (mm)

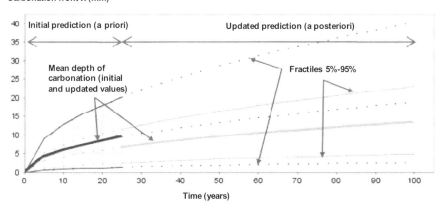

Figure 15.12. *Initial and updated evolutions of the depth of carbonation (mean value and fractiles at 5% and 95%)*

The use of techniques such as Bayesian networks, amongst others, is a useful way to consider real data from the field. These tools help to update the knowledge of the system, and then to predict its future state, which is particularly interesting for the aim of optimizing maintenance actions, for example. These techniques are particularly useful when few data are available (meaning the statistical approach is difficult to implement). Nevertheless, despite the usefulness of this type of approach, it must be noted that the model must be as representative as possible in its description of the phenomena modeled in order to be reliable.

15.4. Conclusion

In the global process of the management of maintenance of civil engineering structures, there are opportunities for optimization from the design stage to the extension of the life span, as illustrated by the two previous examples.

Like any complex system evolving with time, there is some uncertainty about the condition of structures and, generally, few experimental data to characterize them as these are costly to obtain and may require costly unavailability periods.

Therefore, numerical modeling of the ageing process is one solution for the structure's owner to estimate the future evolution of their asset. Many uncertainties exist related to the parameters of these models, and reliability approaches can provide richer results for decision-making than purely deterministic approaches. When field data exist, it is possible to update the initial prediction and therefore adapt the planned maintenance schedule. The combination of these different approaches enables the acquisition of a maximum amount of elements that are helpful for the manager in their decision-making process.

15.5. Bibliography

[AUG 09] AUGE L., CORNISH-BOWDEN I., GERARD B., FRENETTE R., "Exploitation d'une concession autoroutière à fortes exigences de performances: maîtrise du coût des risques associés", GC'09 – Cycle de vie des ouvrages: une approche globale. ESTP Cachan, France, 2009.

[BOE 09] BOÉRO J., SCHOEFS F., CAPRA B., ROUXEL N., "Technical management of French harbor structures – part 2: Current practices, needs, experience feedback of owners", Paralia, no. 2, p. 7.1-7.12, 2009.

[CAP 07] CAPRA B., LE DROGO J., WOLFF V., "Reinforced concrete corrosion: application of bayesian networks to the risk management of cooling towers in nuclear plants", CONSEC'07, Tours, France, June 4-6 2007.

[ELL 95] ELLINGWOOD B.R., MORI Y., "Reliability-based service life assessment of concrete structures in nuclear power plants: optimum inspection and repair", 13th International Conference on Structural Mechanics in Reactor Technology (SMiRT 13); Brazil, p. 529-537, August 13-18 1995.

[ENR 98] ENRIGHT M.P., FRANGOPOL D.M., "Probabilistic analysis of resistance degradation of reinforced concrete bridge beams under corrosion", Engineering Structures, 20(11), p. 960-71, 1998.

[GUE 99] GUEDES-SOARES C., GARBATOV Y., "Reliability of maintained, corrosion protected plates subjected to non-linear corrosion and compressive loads", Marine Structures, 12, p. 425-445, 1999.

[HAG 91] HAGEN O., TVEDT L., "Vector process out-crossing as parallel system sensitivity measure", Journal of Engineering Mechanics, 117(10), p. 2201-2220, 1991.

[KAL 04] KALLENA M.J., VAN NOORTWIJK J.M., "Optimal maintenance decisions under imperfect inspection", Reliability Engineering and System Safety, p. 1-9, 2004.

[LAN 05] LANNOY A., PROCACCIA H., Evaluation et maîtrise du vieillissement industriel, TEC & DOC, Lavoisier, Paris, France, 2005.

[MOR 01] MORI Y., NONAKA M., "LRFD for assessment of deteriorating existing structures", Structural Safety, 23, p. 297-313, 2001.

[NAK 02] NAKANISHI S., NAKAVASU H., "Reliability design of structural system with cost effectiveness during life cycle", *Comp. and Industrial Engrg*, 42, p. 447-456, 2002.

[SCH 91] SCHALL G., FABER M., RACKWITZ R., "The ergodicity assumption for sea states in the reliability assessment of offshore structures", *J. Offshore Mechanics and Arctic Engineering*, 113(3), p. 241-246, 1991.

Conclusion

This book has described and illustrated a number of different methods to assess the reliability, and to improve the prediction of lifetime and the management of civil engineering structures in an uncertain context. It has supplied answers to questions such as:

– how can the most likely failures and the most critical failure scenarios, which could optionally be the basis of risk analysis, be highlighted?

– how can uncertain data, describing the geotechnical characteristics of materials, be represented and used?

– what are the consequences of heterogeneity and variability for structural safety?

– how can the reliability or durability of a system be quantified?

– how can information gained over time be used to update reliability calculations?

– how can a policy of inspection and maintenance be optimized?

Part 1 presented methods of qualitative assessment for structural safety. In an engineering context, these methods allow us to analyze a system, its failure modes, and to model the failure scenarios in order to evaluate their criticality. In this first section, the authors stressed the advantages and limitations of the different methods, and then presented an application of the methods, assessing the criticality of the various scenarios for a hydro civil engineering work.

Part 2 showed how to use available data to describe their heterogeneity and variability. It dealt with the characterization of uncertainty in geotechnical data. This

Chapter written by Julien BAROTH, Franck SCHOEFS and Denys BREYSSE.

part of the book offers a complete set of methods: from the identification of sources of uncertainty, to the classification of data and its statistical representation, through to the modeling of these data. Estimates related to material variability (both average and characteristic values) are provided in this section and, to illustrate the points, a geostatistical study for urban soils (Chapter 5) and another for a shallow footing (Chapter 6), for which reliability aspects are considered, are presented.

Part 3 presented another class of methods for calculating the reliability, called "response surfaces" because the mechanical response of a system, usually not explicit, is approximated by a meta-model, often reduced to an explicit analytical polynomial function. This group of methods has been the subject of recent developments, which are applied in this section to examples of a truss structure, and then to the ossature of a building over several floors. Response surfaces often seem appropriate for problems with a low number of random variables (M <20), but recent methods using polynomial chaos are also able to resolve large stochastic problems (M ~ 50–100), at a reasonable computational cost (N <1000).

Part 4 of the book outlined the problems of time-dependent reliability through a number of methods. The aggregation and unification of data was applied in Part Four to assess the evacuation time required to leave a building on fire (Chapter 9). Then, Bayesian methods and the "PHI2" method, recently developed [AND 04], [SUD 08a] (Chapter 10), as well as Markov Chains Monte Carlo (Chapter 11) were presented. The main applications demonstrated in this part of the book are a serial system, in the case of poor and censored data, a truss structure, and a containment building for a nuclear power plant (Chapter 12).

Finally, Part 5 described maintenance optimization using reliability methods, including a presentation of the concepts of maintenance and lifecycle costs of a system. Cost models for the maintenance of components and systems were defined in order to allow the selection of an optimal maintenance policy. Applications demonstrated in this part of the book cover several issues related to the corrosion of reinforced concrete (a pre-stressed beam and cooling tower, for example).

Users of the methods presented should remain cautious: the result of any study are highly dependent on assumptions made and models used (whether physical, mechanical or probabilistic). Readers should keep in mind the following questions:

– is the problem well-posed and the system being studied well defined, and analyzed by structural and functional approaches? An analysis of a system makes sense only for the problem being solved, especially in the context of a multicriteria analysis. There is not one single unique definition of components and their relationships (see Parts 1 and 5, in particular);

– what is the domain of validity of models, and how representative is the data? Questions about the relevance of statistical data and models must always be raised. It is always advisable to use some kind of probabilistic reasoning; however, the approaches described here (whether or not they use probability theory) are not always applicable.

The regulatory framework for the design of structures (such as Eurocode) is framed on a semi-probabilistic basis, and can take into account various uncertainties. The calculation rules proposed, however, may be insufficient when the variability of the materials is a key parameter governing the response, and the use of a more sophisticated modeling is then necessary.

Beyond the applications proposed in this book, there is also ongoing research being conducted in various fields, such as:

– the optimization of reliability based structural design;

– the optimizing of campaigns of inspection or repair;

– the formalization of expert judgments;

– building databases (of material properties, geometry, stresses, boundary conditions, etc.) and updated models which progressively integrate new data; and

– the prediction of the reliability of ageing structures.

Prediction is difficult, requiring a number of assumptions and extrapolations to be made. This last point is crucial, since these methods are often the only ones to offer a theoretical framework for evaluating the performance of existing structures, which is a major challenge for the development of our societies.

The authors are convinced that the methods presented in this book are applicable to any complex mechanical system in an uncertain environment, although the examples presented are limited to the field of civil engineering (whether for nuclear industry, oil industry, or dam building applications).

List of Symbols

Indices

i	Index of a phenomenon or component
j, k	Index of a scenario and function
l	Index of a data or proof

Data

$E_i(t)$	State of a component i at time t
Ph_i^j	Phenomenon i of scenario j; *a priori*, a phenomenon i corresponding to a scenario j is denoted as Ph_i
Sc_j	Scenario j
X, Y	Two variables, where $\overline{X}$ and $\overline{Y}$ are their norms
x, y	Realizations of X and Y, where $\overline{x}$ and $\overline{y}$ are their norms
f_{perf}	Level of performance

Physical model

$\boldsymbol{x} = \{x^i, i=1,...,P\}$	Deterministic vector – input parameters
$\boldsymbol{d}$	Vector of parameters of a physical model (which are not random variables)
$\boldsymbol{d'}$	Vector of parameters of criterion
$\mathbf{M}(\boldsymbol{x}, \boldsymbol{d})$	Physical model
$\underline{y} = \mathbf{M}(\boldsymbol{x}, \boldsymbol{d})$	Response of a physical model
$z = g(\boldsymbol{x}, \boldsymbol{d})$	Criterion of interest

Quantification in time

t	Instant, date
t_0 , t_f	Beginning or end of the study
da_i^j	Beginning of a phenomenon i of a scenario j, Ph_i^j
da^j , dr^j	Beginning of the realization of a scenario Sc_j
dr_i^j	Realization of a phenomenon i of a scenario j, Ph_i^j
df_i^j	End of a phenomenon i of a scenario j , Ph_i^j
Du_i^j	Duration of a phenomenon i of a scenario j , Ph_i^j
Du^j	Duration of a scenario Sc_j

Quantification of criticality

G^j , Cr^j	Gravity, criticality of a scenario Sc_j
C_M , C_F	Total cost of maintenance and mean cost of failure, respectively
C_{PM} , C_{INS} , C_{REP}	Costs of preventive maintenance, inspection and repair, respectively

Quality and fusion of data

Θ	Frame of discernment
θ	Event of Θ
m	Belief mass
$\mu_{FS}(x)$	Function of belief mass
FS_l	Fuzzy subset of Θ, concerning data l
$Ker(FS)$	Kernel of the fuzzy subset FS_l
$Supp(FS)$	Support of the fuzzy subset FS_l
$h(FS)$, $|FS|$	Height, cardinality of the fuzzy subset FS_l
FS_α	α-cut f the fuzzy subset FS_l
$\min_{FS}$, $\max_{FS}$	Min, max of the fuzzy subset FS_l
$C_f(\theta)$	Consensus of θ
$Bel(\theta)$	Belief in θ
$Pl(\theta)$	Plausibility of θ
$P_S(\theta)$	Smets probability of θ
$Ind(\theta)$	Indicators of quality of θ , with $Ind(\theta)=\{Bel(\theta);P_S(\theta);Pl(\theta)\}$

Probabilities

X	Random variable (r.v.)
$X=\{X^i, i=1,..,M\}$	Random vector
$F_X(x)$	Cumulative distribution function of the random variable X
$F_X(x)$	Cumulative distribution function of the vector X
$f_X(x)$	Probability density function of the random variable X
$f_X(x)$	Joint probability density function of the random vector X
$\mathbf{E}[.]$	Expectation

$\text{Var}[.]$	Variance
$\text{Cov}[.,.]$	Covariance
$\rho_{X^i,X^j}, \rho_{ij}$	Coefficient of correlation between X^i and X^j
$\mathbf{C}, \mathbf{R}$	Matrices of covariance, of correlation
$\text{P}(A)$	Probability of an event A
P_f	Probability of failure
$R = 1 - P_f$	Reliability
μ, μ_X	Expectation of X (mean)
σ, σ_X	Standard deviation of X
σ^2, σ_X^2	Variance of X
cv_X	Coefficient of variation of X
δ, δ_X	Skewness of X
κ, κ_X	Kurtosis of X
μ'_n, μ_n	n-order statistical moment, zero-mean n-order moment
$\mathbf{1}_A(x)$	Characteristic function of the condition A
$\mathbf{N}(\mu, \sigma^2)$	Normal (or Gaussian) distribution of mean μ and of standard deviation σ
$\mathbf{N}(0,1)$	Standard normal distribution (zero-mean and unit-variance)
$\varphi(x)$	Probability density function of the standard normal variable
$\Phi(x)$	Cumulative distribution function of the standard normal variable
ξ	Standard random variable
$\mathbf{LN}(\lambda, \zeta)$	Lognormal distribution with parameters λ and ζ
$\mathbf{E}(\lambda)$	Exponential distribution of parameter λ
$\mathbf{W}(a, \beta, \eta)$	Weibull distribution with parameters a, β, η
$\mathbf{U}(a,b)$	Uniform distribution of bounds a, b
$\mathbf{B}(p)$	Bernoulli distribution of parameter p
$\Gamma(\alpha, k)$	Distribution Gamma of parameters α, k
$f_{Y|X}(x,y)$	Conditional probability density function
$F_{Y|X}(x,y)$	Conditional cumulative distribution function
$\mathbf{E}[Y \mid X]$	Conditional expectation
$\text{Var}[Y \mid X]$	Conditional variance

Some useful formulas

$$F_X(x) = \int_{-\infty}^{x} f_X(x)\, dx \ ;$$

$$\mathbf{E}[X] = \int_{\mathbb{R}} x\, f_X(x)\, dx \ ;$$

$$\mathbf{Var}[X] = \int_{\mathbb{R}} (x - \mu)^2\, f_X(x)\, dx \ ;$$

$$\mu_n = \mathbf{E}\big[(X - \mu)^n\big] = \int_{\mathbb{R}} (x - \mu)^n\, f_X(x)\, dx \ ;$$

$$cv_X = \sigma_X / \mu_X \ ; \quad \delta_X = \mathbf{E}\big[(X - \mu)^3\big]\big/ \sigma^3 \ ;$$

$$\kappa_X = \mathbf{E}\big[(X - \mu)^4\big]\big/ \sigma^4 \ ;$$

$$\mathbf{Var}[Y] = \mathbf{Var}\big[\mathbf{E}[Y \mid X]\big] + \mathbf{E}\big[\mathbf{Var}[Y \mid X]\big] \ ;$$

$$\mathbf{C} = \mathbf{Cov}[\mathbf{X},\mathbf{X}] \ ;$$

$$\mathbf{C}_{ij} = \mathbf{Cov}\big[X_i, X_j\big]$$

Statistics

N	Sample size
$\{x_1, .., x_N\}$	Sample
$\boldsymbol{\theta} = \{\theta_q,\ q=1,..,N\}$	Parameters of the probabilistic model
$\hat{\theta}$	Parameter estimator
$\bar{x}$	Empirical mean of the sample $\{x_1, .., x_N\}$
$s^2,\ s'^2$	Empirical biased and non-biased variance

Structural reliability

$g(X)$	Performance function
$\mathbf{D}_f,\ \mathbf{D}_s$	Domains of failure and safety

$X = T(\underline{\xi})$	Isoprobabilistic transformation
P_f ,	Exact probability of failure
$\hat{P}_f$	Estimated probability of failure
$P_{f,FORM}$, $P_{f,SORM}$	Probabilities of failure approximated by FORM and SORM, respectively
β_{FORM}	Hasofer–Lind reliability index (FORM)
β_{SORM}	Generalized reliability index (after SORM calculation)
$P*$	Design point
$X*$	Coordinates of the design point (physical space)
$\underline{\xi}^*$	Coordinates of the design point (standard space)
α	Vector of direction cosine of the design point
α_i^2	Importance factor of the *i-th* variable

Common acronyms

Maintenance strategies

ABAO	As Bad As Old
AGAN	As Good As New
ALARP	As Low As Reasonably Practicable
BTO	Better than Old
CM	Corrective Maintenance
NBU/NWU	New is Better/Worse than Used
PM	Preventive Maintenance
WTO	Worse Than Old
Mean times	
MDT	Mean Down Time
MTTF	Mean Time To Failure

MTBF	Mean Time Between Failures
MTTR	Mean Time To Repair
MUT	Mean Up Time
MWT	Mean Waiting Time

Other acronyms

AIC	Akaike Information criterion
AIS	Adaptative Importance Sampling
ARA	Arithmetic Reduction of Age
BIC	Bayesian Information Criterion
PC	Polynomial chaos
SLS	Service Limit State
ULS	Ultimate Limit State
EM	Expectation Maximization
FMEA	Failure Modes and Effects Analysis
FORM	First Order Reliability Method
LHS	Latin Hypercube Sampling
MCMC	Markov Chain Monte Carlo
NDT	Non-Destructive Testing (Device)
PCA	Principal Component Analysis
PCP	Probability of Crack Presence
PFA	Probability of False Alarm

PDRS	Polynomial Development Response Surface
PO	Planned Operation
PoD	Probability of Detection
PRA	Preliminary Risk Assessment
QRA	Qualitative Risk Analysis
RBI	Risk Based Inspection
RCM	Reliability Centered Maintenance
ROC	Receiver Operating Characteristic
SEM	Stochastic Expectation Maximization
SORM	Second Order Reliability Method
SR	Response surface

List of Authors

Julien BAROTH
IUT1/ Laboratoire 3SR
University of Grenoble 1
France

Marc BERVEILLER
EDF R&D
Moret-sur-Loing
France

Géraud BLATMAN
EDF R&D
Moret-sur-Loing
France

Daniel BOISSIER
Polytech' Clermont-Ferrand/LaMI
Blaise Pascal University (Clermont II)
France

Denys BREYSSE
GHYMAC Laboratory
Bordeaux University 1
France

Bruno CAPRA
OXAND
Avon
France

Gilles CELEUX
INRIA
Saclay-Île-de-France
France

Alaa CHATEAUNEUF
Polytech' Clermont-Ferrand/LaMI
Blaise Pascal University (Clermont II)
France

Frédéric DUPRAT
LMDC Laboratory
INSA of Toulouse
France

Antoine MARACHE
GHYMAC Laboratory
Bordeaux University 1
France

Laurent PEYRAS
Cemagref
Aix-en-Provence
France

Franck SCHOEFS
GeM
University of Nantes
France

Bruno SUDRET
Phimeca Engineering
Paris
France

Aurélie TALON
Polytech' Clermont-Ferrand/LaMI
Blaise Pascal University (Clermont II)
France

Index

D, E

F, G